# THE END
# OF TIME

## ALSO BY ANTHONY AVENI

*Ancient Astronomers*

*Behind the Crystal Ball: Magic, Science and
the Occult from Antiquity Through the New Age*

*Between the Lines: The Mystery of the
Giant Ground Drawings of Ancient Nasca, Peru*

*The Book of the Year: A Brief History of Our Seasonal Holidays*

*Conversing with the Planets: How Science and Myth Invented the Cosmos*

*Empires of Time: Calendars, Clocks and Cultures*

*The First Americans: Where They Came From and Who They Became*

*Foundations of New World Cultural Astronomy*

*The Madrid Codex: New Approaches to
Understanding an Ancient Maya Manuscript* (with G. Vail)

*Nasca: Eighth Wonder of the World*

*Skywatchers: A Revised and Updated
Version of Skywatchers of Ancient Mexico*

*Stairways to the Stars: Skywatching in Three Great Ancient Cultures*

*Uncommon Sense:
Understanding Nature's Truths Across Time and Culture*

# THE END OF TIME

## THE MAYA MYSTERY OF

# 2012

### ANTHONY AVENI

UNIVERSITY PRESS OF COLORADO

For Dylan

© 2009 by Anthony Aveni

Published by the University Press of Colorado
5589 Arapahoe Avenue, Suite 206C
Boulder, Colorado 80303

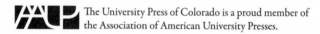 The University Press of Colorado is a proud member of
the Association of American University Presses.

The University Press of Colorado is a cooperative publishing enterprise supported,
in part, by Adams State College, Colorado State University, Fort Lewis College,
Mesa State College, Metropolitan State College of Denver, University of Colorado,
University of Northern Colorado, and Western State College of Colorado.

∞ The paper used in this publication meets the minimum requirements of the
American National Standard for Information Sciences—Permanence of Paper for
Printed Library Materials. ANSI Z39.48-1992

Library of Congress Cataloging-in-Publication Data

Aveni, Anthony F.
 The end of time : the Maya mystery of 2012 / Anthony Aveni.
  p. cm.
 Includes bibliographical references and index.
 ISBN 978-0-87081-961-2 (pbk. : alk. paper)  1. End of the world (Astronomy)
2. Maya calendar. I. Title.
 QB638.8.A94 2009
 001.9—dc22

                                                            2009023097

Design by Daniel Pratt

18  17  16  15  14  13  12  11  10  09        10  9  8  7  6  5  4  3  2  1

# CONTENTS

# FOREWORD

Several decades ago when I was a graduate student, I asked a professor about the possibility that a Maya building was astronomically oriented. His response was to scoff that there were so many stars in the sky that it was inevitable that a building would be oriented to at least one of them. (Apparently, he was of the mind that there are no stupid questions, only stupid people asking questions.) But that was then and this is now, and thankfully such queries are no longer treated so dismissively. For this, we owe a tremendous debt to the careful cross-cultural observations and painstaking astronomical measurements of Anthony F. Aveni, who has pioneered the development of archaeoastronomy (or cultural astronomy) into a highly respected science worldwide.

In *The End of Time: The Maya Mystery of 2012*, Aveni treats us to a thoughtful analysis of the burgeoning pseudo-theories attached to the closing date of the Maya calendar in December of that year. As his career and many awards attest, Aveni is a lifelong teacher, and in *The End of Time* he teaches all of us—astronomers, Mayanists, and the general public—about the complexities of how the Maya arrived at this date and what it might mean for the modern world. He debunks the outlandish claims of future catastrophes by systematically evaluating scientific data: he walks the walk and "does the math." But beyond that, he is interested in what this current fascination with the 2012 ending date says about *us*: why are we—especially twenty-first-century Americans—so preoccupied with divining its meaning?

Aveni reviews philosophical and intellectual trends in Western thought, going back to Classical and Biblical traditions, especially Gnosticism, and apocalyptic histories, but he also involves long-standing Euro-American romanticized images of indigenous inhabitants of the Americas and the native wisdom they embodied. Americans today have a puzzling anti-science streak that is manifest in many ways, including rejection of evolutionary theory and an uncritical acceptance of bizarre notions about lost continents, ancient astronauts, Y2K disasters, planetary conjunctions, and all kinds of similar hokum, much of which is purveyed by the Internet and Hollywood. (This bipolar tendency was certainly evident in the lead-up to the 2008 national elections: at the same time that voters railed against "the liberal professoriate" in our nation's universities, they were simultaneously demanding greater access to [read "cheaper"] college educations for their children. Who do they think are teaching these children?)

Aveni is generous in his assessment of why the Y12 phenomenon has gotten traction with today's populace: it is less about global cataclysm and more about a rejection of Western cultural imperialism and a desire, in effect, to get in touch with our kinder, gentler selves. He is not one of the "haughty, exclusive establishment" scientists that bash the views of the non-cognoscenti. Instead, he arrives at an understanding that Y12 aficionados share with the fringe theorists a deep concern about the origins of humans and civilizations and their ultimate fate.

In much of the content of the pseudoscientific theories that have come and gone in public consciousness is a sense that the answers to these puzzles lie in "secret knowledge" of ancient civilizations or are encoded in certain rhythmic repetitions in the natural world, such as sunspot cycles or acupuncture points. The Maya had their secret knowledge, to be sure, but this knowledge was about the gods who carried the burdens of cosmic time.

Like many prescientific peoples, the Maya believed that time moved in cycles. Lots of cycles. Cycles of 20 days, cycles of 65 days, of 260 days, 365 days, 20 years, 52 years, 260 years, 400 years, and on and on. Generally, cyclical time is ritual time and mythical time. But in publicly celebrating the end of one cycle and the beginning of a new one, the terrifying possibility always lurks that the new cycle will not actually restart. In the face of this uncertainty, cyclical time has the advantage of being "controllable": a Maya king and his calendar priests can command labor and tribute and sacrifice on the part of the masses to appease the gods and ensure that that the cycle will renew . . . and when it does, they can triumphantly proclaim their control over the gods of time as the sun rises once again to start a new day and new cycle. Such esoteric knowledge about the timing of cycles' endings and rains' startings and eclipses' occurrences was kept secret from the people in order to maintain the mystical power of the sacred king, or *k'ul ajaw*. Like his royal ancestors, the divine king, who appeared before his subjects as the Sun God and the Maize God on ceremonial occasions, undergirded his absolute authority by seeming to "control" the cosmos, the rains, the maize crop . . . and of course the people, who were held in thrall to this world view.

Along with these multiple, ongoing cycles of time, the Maya believed—as did the peoples of many ancient civilizations—that there were multiple creations of the world, animals, and humans. According to the highland K'iche' Maya creation myth *Popol Vuh*, we are nearing the end of the fourth of these creations. The first three began with the gods' unsuccessful attempts to create humans who would "pray to them and keep the days." That is, the gods wanted humans to be able to speak intelligibly and observe proper rituals on the proper days of the calendar. Humans were unable to do this until they were created of maize, thereby ensuring that this fourth creation of the universe was a success.

According to the Maya, this creation and calendar cycle, a "great cycle" of 5,126 years, began on August 13, 3114 BC and will end on December 21, AD 2012. The Maya dated important events in the lives of their dynasties and cities by means of what archaeologists call the Long Count. The Long Count is truly looooong: it allows the Maya to situate any important event by counting the number of elapsed days in multiple intersecting cycles since this starting date of the present creation. Imagine if we had to date all our letters or e-mails or blog entries by counting how many days had elapsed since January 1, AD 1! (In fact, Julian calendars used by astronomers do this very thing, but today most of us use the more convenient Gregorian calendar with its units—cycles!—of weeks and months.)

But it is difficult, as Aveni and others (including me) have noted, to understand why August of 3114 BC and December of AD 2012 are the termini of the Long Count. We are 99.99 percent certain that the Long Count was not invented in 3114, which means that we have to find a plausible date when it did begin. Did later Maya select a beginning date (in 3114 BC) and calculate through numerous cycles an ending in 2012? Or did they calculate an ending date in 2012 and work backward to retrodict a starting year in 3114? These are not easy questions to answer, and as Aveni laments repeatedly and justifiably, we are severely limited in our ability to investigate these and related issues by the abysmal lack of textual evidence. We do not know if the Maya wrote about these matters, but if they did, the writings on bark paper did not survive the centuries in tropical climes—or the zealotry of the early Spanish priests who burned them.

So on December 21, AD 2012, as the old Maya calendar cycle ends, a new one will start all over again. The archaeologists' notation 13.0.0.0.0—the day of completion of thirteen Maya 400-year baktuns—is also 0.0.0.0.0, the first day of the new baktun. There is no reason to believe that our world and its humans, Maya or non-

Maya, will cease to exist on this day, because Maya priests and shamans and "daykeepers" have been faithfully "keeping the days" for millennia, up into the twenty-first century. Thus, the Maya do not tell us of our ultimate fate, but as *The End of Time* makes clear, they do remind us what we have in common with other people in other times and places.

**PRUDENCE M. RICE**
Southern Illinois University Carbondale

# PREFACE

How will it end—cosmic collision, global climate change, nuclear holocaust? When will it end—in billions or millions of years, in a handful of generations—or in only a few years? What makes us think there *ever* will be an end to the world as we know it? Maybe the world is eternal? Big questions that are right up there with the search for the meaning of life.

Who knows? How can anyone know? Did our ancestors know? We certainly seem primed to know. I did a little survey of end-of-the-world predictions since the 9/11 destruction of the World Trade Center. There were a dozen listed for 2006 alone, including two that portended the second coming of Christ (June 6, December 17), one the Islamic Armageddon (August 22), two a nuclear war (September 8–9, September 12), one a collision with a comet (May 25), and one a great earthquake (January 25). Five predictions were non-specific as to both cause and date.

Other recent prophecies predict positive events—global awakening, hyperspatial breakthrough, sudden evolution of *Homo sapiens* into non-corporeal beings, and even the return of alien caretakers to assist us—or events more negative in nature, asteroid collision, nuclear war, reversal of the earth's magnetic field, world blackouts because of the oil crisis, return of the apostle Peter and the destruction of Rome, and, or course, the reappearance of alien caretakers bent on enslaving us.

*The End of Time: The Maya Mystery of 2012* probes these fascinating questions, especially the theory that advanced knowledge about the ultimate outcome of humanity and planet earth is secretly encoded in ancient documents that have been passed down through the ages—documents interpretable only by those capable of acquiring higher knowledge.

The study of "last things" has a name of its own. It's called *eschatology* (from the Greek word *eschatos*, meaning furthest in time). Eschatology divides sharply into two doctrines based on how time is understood. The *mythic* doctrine, widespread in many cultures, sees humanity immersed in a struggle between the forces of order and chaos. People derive meaning from the rituals they conduct to see the world through its impending destruction and the creation of a new world. In most versions, mythic time is cyclic. Destruction and renewal happen over and over again, endlessly. *Historical* eschatology, derived from Judeo-Christianity, is based on a linear understanding of time. The world will suffer singular destruction because of humanity's violation of the laws of God, but existence in the eternal world to follow is possible provided we seek salvation and redemption before time's end. The contemporary Christian version of what awaits us is heavily laden with apocalyptic overtones—the idea that God will intervene violently and suddenly at a preordained moment in time.

The mythic idea that world ages, marked by beginnings and endings of great calendrical cycles, are preordained in the stars belongs to both doctrines and it is widespread and deeply rooted in Western history. This idea has enjoyed a resurgence in American pop culture, especially since the revolutionary 1960s. It approaches a frenetic crescendo in recent prophecies about the impending end of the world in 2012, thought by many to emanate from ancient Maya wisdom. Some prophets say the end of the Maya Long Count cycle (one of many ways the ancient Maya reckoned time) on the winter solstice of that year will be attended either by an apocalyptic doomsday or by a sublime ascent to a higher consciousness. Whether doom or bliss awaits us all depends on which visionary you listen to.

Why the Maya? How does such a remote culture manage to acquire such a powerful hold on so many of us? Who were these people? We know they are alive today, but what do we know about their ancient calendar, their astronomy, their cosmology, and especially their ideas about the creation and destruction of the world? Was the great cycle of the precession of the equinoxes—the wobbling of the earth on its axis—part of the Maya plan, as some suggest? These are a few of the questions we will probe in *The End of Time: The Maya Mystery of 2012*. I think they are linked to even more basic questions, Why do we reach into the deep past of another culture to acquire truths about ourselves? What compels contemporary Anglo-American societies to think that the message of the ancient Maya is intended for us? Why are many of us entranced enough by the Maya mystique to travel vast distances to the ancient ruins at specially designated times to gain access to the power point of Maya prophecy?

Who am I to tackle such profound questions about star-fixed Maya determinism? I was trained originally in astronomy and I have spent most of my life studying Maya calendars. As a result, I have had the opportunity to field lots of questions about Maya astronomy and cosmology. I first began receiving inquiries concerning 2012 about ten years ago. At this writing they are too numerous to respond to.

I know it is not fashionable for academics to write popular books, but in this instance an e-mail correspondence with a young high-school student pushed me over the line. Dylan was worried, but he was also intensely curious about all the 2012 hype. I could not resist my natural inclination to teach—my true calling in life. But a serious teacher should also be a good listener and a good learner, skills I have tried to practice in putting this work together. Above all, what I learned about 2012 is that Will Shakespeare may have had it right: the real truth may lie more in ourselves than in our stars.

## ACKNOWLEDGMENTS

To my friends and colleagues with a wide variety of perspectives on issues raised in this book for their willingness to discuss them with me: Gary Baddeley, Harvey and Victoria Bricker, Robert Garland, Joscelyn Godwin, John Justeson, Tim Knowlton, Susan Milbrath, Mary Miller, William Peck, Prudence Rice, Barry Shain, David Stuart, Gabrielle Vail, Mark Van Stone, Chris Vecsey, and Belisa Vranich.

And to those in the production line: Darrin Pratt, Dan Pratt, Laura Furney, Beth Svinarich, and the staff at University Press of Colorado on this, our seventh project together; Samantha Newmark (my student for three years); Diane Janney (my extraordinarily able assistant for more than a decade); Faith Hamlin (my thoughtful agent for two); and Lorraine Aveni (my unremitting muse for five).

# THE END OF TIME

# INTRODUCTION:
# HOW DYLAN GOT ME STARTED

On December 21, 2012 (or December 23, 2012, depending on how you align their ancient calendar with ours), the odometer of ancient Maya timekeeping known as the Long Count will revert to zero and the cyclic tally of 1,872,000 days (5,125.3661 years) will start all over again. When I first became attracted to Maya studies over forty years ago I could not possibly have imagined that I would write a book about this event. Blame Dylan.

Three years ago I began receiving e-mails from a troubled Canadian high-school student, Dylan Aucoin, from Dartmouth, Nova Scotia. He had been reading Web articles about the end of the world that would supposedly fulfill the Maya prophecy about what might accompany the Long Count's great turnover in 2012—or Y12 as I have come to call it. Dylan confided to me that he was worried—at times even horrified—by the predictions he had come across: apocalypse, holocaust, world destruction. After encountering one particularly frightening doomsday article, Dylan asked me:

"Is there anything to fear about 2012 and the New Age ideas of destruction and consciousness shifting? I thought I had it all figured out but this article has brought it back like gangbusters. I ask you, is it worth fretting about? Is there really any validity?"

At first I thought he was putting me on but there was a sense of urgency in the tone of Dylan's words that spurred me to respond. We began a joint reading program and conversation. I was impressed with Dylan's motivation to investigate things for himself, a quality I admire when I see it in my students. Dylan told me of an encounter with an old man who came into the Blockbuster store where he worked near Halifax, Nova Scotia. In casual conversation the old man told him his whole family was brilliant, with IQs topping 150 and that he himself was a member of an ancient mystical order that derives its truths from psychic teachings. You had to have "the gift" to know the deepest truths, he told Dylan. Although skeptical, Dylan listened to the end-of-the-world prophecies the old psychic spouted. Dylan told me that whenever someone discusses anything with such passion he gets motivated to investigate the subject. A film buff, Dylan wondered why many famous celebrities bought into the idea of psychic phenomena, especially as they relate to great world transformations—from actress Shirley MacLaine to baseball star Darren Daulton. We discussed actor Billy Bob Thornton's film *The Gift*, about his mother's life as a psychic, and the Jim Carrey film *The Number 23*, about a man who becomes obsessed with the way everything numerological in his life, like the number of letters in his name, seminal dates—and the sum of 20 + 1 + 2 in 2012— added up to 23. Coincidence?

Dylan and I wove our way through a panoply of Internet sites on 2012 and a pile of 2012 texts and articles, many of which he recommended to me. Meanwhile, I tried to fill him in on what I knew about the Maya calendar and cosmically provoked disasters.

After a year of correspondence I could see the spark of natural curiosity inherent in the now seventeen-year-old really catch fire. Dylan wrote me a neat little passage on his acquired view of skepticism. Skepticism is labeled by believers of bizarre theories as outright rejection and arrogant self-appointment—"a buzzkill" as he put it. "But skeptics aren't just stubborn people who don't want to believe. Good skeptics want to listen. They just want evidence—and that's where a lot of claims made by psychic prophets fall short." Wise words from a man so young. "I wish I'd been exposed to critical thinking earlier in my life," he added. Dylan was getting the picture. Real skepticism is about self-criticism, questioning, and the passionate search for evidence. Never be satisfied. Dylan's words inspired me to share the skeptical view with a wider audience.

As the fateful day draws nearer I continue to receive more e-mails and I feel duty-bound to field more questions about the meaning of 2012 at lectures and conferences. The head of an Ask the Experts site based at the University of London's Institute of Education asked me to do a little piece on calendar cycles and world ages to shed some light on what he called "the vexed question of December 2012. It keeps coming up in the schools we work with," he said, and seems to be "surrounded by hype." Could I "demystify" it?

Spurred on especially by my probing of books, articles, and Web-based material such as Beyond 2012: Catastrophe or Ecstasy?, thanks to Dylan and others, I began to understand the concern. Many prophecies are filled with frightening forecasts for the near future. To list just a few:

- The great Maya lord will make everything die.
- The world as we know it will come to an end.
- Damaging sunspots will reach their peak.
- The Cosmic Shaman of Galactic culture offers us clues for healing the planet which will be destroyed if we don't act now.
- The solar system will enter an energetically hostile part of the galaxy.
- Mass extinction will take place.
- Yellowstone will explode.

- The earth's magnetic poles will reverse.
- We may get sucked into a black hole.

Dylan was right. These end-of-the-world scenarios *are* terrifying.

Now that you've read a page or two of this book, you are probably thinking that I am going to pooh-pooh all those doomsday predictions, playing the archetypal academic role of debunker. In every scientific controversy there is always an "expert" who bashes the outsiders, those unheralded independent thinkers who just might be glimpsing truths unrecognized by the haughty, exclusive establishment. I am not a dismissive academic. I always take my listeners and my readers seriously—I read what they write. When I wrote *Behind the Crystal Ball*, a book about the history of magic and occult beliefs, I interviewed astrologers, palm readers, and channelers. Rather than focusing on debunking non-mainstream beliefs, and well aware that many of these prognosticators truly believed in the doctrines they promoted, my interest lay in why people believed in magic and how occult beliefs changed with the times. My agnostic take on occult behavior netted me more than a few negative reviews.

I am interested in questions about ourselves, and I write when I think I have something to say about our culture that has not already been said. In the vast quantity of material on 2012 that I have studied since Dylan kick-started me on this project, I do not think the part of the story about *us* has really been articulated. Why has 2012 become such a big deal in contemporary mainstream culture? What is it about the ancient Maya and the end point of their calendar that makes so many of us take notice? Why are so many people today convinced the Maya message is meant for us? Do we have any clues to help us understand what the great cyclic turnover meant to them? What prophecies did it portend? These, in my opinion, are the most interesting questions to ask, and I think their answers can help us understand what lies behind the current Y12 mania.

In the next chapter, "What's in Store: A User's Guide to 2012 Maya Prophecies," I will survey the major works and the cast of characters that have spurred most of the attention on the 2012 event. These include Jose Argüelles, who focused on calendar convergences in the premillennial 1980s. His "Harmonic Convergence," the coming together of sub-cycles of the Mesoamerican calendar in 1987, was the warm-up act to 2012. We are at a point just twenty-six years short of a major galactic synchronization, he theorized. Either we shift gears right now or we will miss the opportunity. John Major Jenkins foresees a great galactic alignment attending 2012—the winter solstice sun crossing the center of the Milky Way. He believes that he has found evidence to support his theory that when the ancient Maya invented the Long Count 2,000 years ago, they deliberately geared it to this alignment. What is more, he implies, the great astronomical event will provide the signal that will usher in a new consciousness.

Whereas Argüelles and Jenkins base their scientific-sounding theories of time's big overturn on astronomical and calendrical calculations, other sages, like Carl-Johan Calleman and Daniel Pinchbeck, claim to acquire their knowledge about Y12 via special insight—either by associating with Maya shamans or by becoming shamans themselves. These "adepts" contend that their personal odysseys and transformations have endowed them with special powers of insight into the meaning of the end of the Maya creation. Finally, synthesizers, such as Lawrence Joseph, draw from both the scientific and religious outlooks, arriving at a fear-wracked prognosis. They try to show (to quote one book jacket) "why the year 2012 will likely be more tumultuous, catastrophic, and quite possibly revelatory than any other year in human history. Nail biters beware . . ."[1]

Once I have laid out what the modern prophets say will happen in 2012, I will focus on the Maya themselves. In Chapter 3, "What We Know about the Maya and Their Ideas about Creation," and Chapter 4, "The Calendar: Jewel of the Maya Crown," I assess the evidence gathered by archaeologists, historians, and epigraphers (those who study Maya hieroglyphs) on the origins of Maya culture and the likely reasons behind the ascent of the Maya of Yucatan to the loftiest peaks of ancient cultural achievement.

We do not refer to the heyday of the Maya as the "Classic period" whimsically. They built great cities, erected colossal pyramids, and developed an advanced system of elaborate syllabic writing with more than a thousand hieroglyphic signs representing different sounds. They devised a system of timekeeping unparalleled in the Old World that included numeration by position and the concept of zero. In the field of astronomy they followed the movement of celestial bodies with uncanny accuracy, carefully tabulating cosmic cycles in painted documents called codices, made out of tree bark, and advertising them on massive stone monuments known as stelae. And they achieved it all with a minimum of technology. No wonder the Maya impress us. They are the Greeks and the Babylonians of the New World.

The Maya have captured our contemporary imagination like no other ancient culture since the Egyptians back in the 1920s, when archaeologists first breached the tomb of King Tutankhamen. From furniture design to women's fashion, from hairdos to horror films, Egyptian relics—above all, their pyramids—evoked a sense of mystery about the past and the secrets it might yet hold. King Tut still draws crowds every time his sarcophagus goes on exhibit.

The Maya are today's Egyptians of the New World as well. Ever since outsiders first set foot in Yucatan nearly 200 years ago, we have thought of the pyramid builders who once lived there as the "mysterious Maya." Enveloped in all the mystery—in ancient codices and carved stelae, as well as in the stories of creation told in the *Popol Vuh* and in the sacred Books of Chilam Balam—are the ideas we will need to confront and explore if we really want to understand what the Maya thought came before and would come after the present Long Count creation epoch.

Because astronomy emerges as one of the most important considerations in so many 2012 end-of-the-world scenarios, we will need to acquire a basic understanding of "The Astronomy behind the Current Maya Creation," the title of Chapter 5. What exactly is the Milky Way and how does it appear to us in the sky? What exactly is the alignment that attracts so much attention? We will also need to investigate sunspot cycles, magnetic field reversals, and the long-term astronomical cycle known as the precession of the equinoxes. All of these phenomena are alleged by one contemporary 2012 prophet or another to lie at the foundation of Maya apocalyptic predictions. If the Maya knew about these things, we must understand how these phenomena operate. I take particular interest in astronomical issues because I was trained in that field before I ever laid eyes on the Maya. Applied to Maya studies, archaeoastronomy (or cultural astronomy), a field I helped establish, examines the unwritten record in broken bits of Maya archaeological artifacts and standing structures as well as the written record and images in native texts. Its goal is to understand how the Maya used material media to express their knowledge of the sky. As we will discover, the cosmos really *did* play a huge role in shaping Maya thought and action.

In Chapter 6, "What Goes Around: Other Ends of Time," we will learn that scenarios regarding "last things" around the world have a lot in common. In particular, I will trace the Western Judeo-Christian apocalyptic idea of time's end, highlighted in controversial Biblical texts. We will follow different interpretations of the historical view of time, from the Gnostic philosophy of the early Christian era up to the doorstep of American culture, where it begins to combine with aspects of the mythic view. We will explore some of the tensions that accompany all big cyclic endings of time, including the fin-de-siècle, or end-of-century, events in our own Western calendar.

Anyone who browses the Web or the shelves of bookstores and libraries will discover that America has a particular fascination with Y12. Why is that? My closing Chapter 7, "Only in America," links our contemporary world view with the evolution of thought about the end of the world. Here I will pick up the thread of apocalyptic thinking from the previous chapter and follow it from seventeenth-century Puritan New England through the occult fads of the mid-nineteenth century, the revolutionary 1960s, and the 1970s, when epigraphers cracked the code of the mysterious Maya glyphs. Only then can we explore what really lies behind our love affair with the ancient Maya and the 2012 prophecy's grip on pop culture.

Peering into another culture's view of the world is a little bit like looking through a telescope. When I gaze at that far away Other beyond the glass, I might glimpse a different kind of mathematics, different writing, different astronomy, and different architecture, religion, philosophy, folkways and so on. The closer I look the better understanding I acquire of my own narrow, time-squeezed, contemporary way of comprehending the world around me. The more, too, do I realize that cultural differences can be vast. It is all too easy to yield to the temptation of garbing the Other in our own Western clothing. How many times have I heard the confession from my students that studying the Maya has impacted their lives? That is my goal in *The End of Time: The Maya Mystery of 2012:* to understand what the ancient Maya really had to say about 2012.

# WHAT'S IN STORE? A USER'S GUIDE TO 2012 MAYA PROPHECIES

Metaphysical travel—also called sacred travel—is a burgeoning branch of today's tourist industry. Popular destinations tend to be mysterious places, especially those many of us have difficulty believing could have been constructed by an ordinary human labor force. Egypt's pyramids, Stonehenge, and Machu Picchu all come to mind. "[W]hoever built them built them on places that were already places of power on the earth—the acupuncture points on the earth's body that hold powerful energies," notes the proprietor of Body and Mind Spirit Journey, a travel outfit based in Sedona, Arizona.[1] People who sign up for these journeys of recreational self-discovery say they do it to connect with the unique spiritual energy or higher knowledge they believe they will find at these universal sacred places. For example, metaphysical travelers seek the accumulated wisdom of the advanced civilization of Atlantis, thought by some to have been secretly deposited on the site of the Great Pyramid of Khufu.

To be effective, metaphysical appointments need to be kept on *time*—and you must be in the right place. Historians of religion call them *hierophanies*, or manifestations of the sacred. They can be good, such as the Virgin Mary appearing on the wrinkles of a plate glass window or a weeping statue, or evil, such as the plume of smoke in the shape of a devil that many saw issuing from the destruction of the World Trade Center. Hierophanies can be great crowd pleasers. They invite participation; they evoke a feeling of being connected. Some witnesses feel as if their participation actually helps bring about the event.

One of Mexico's most popular destinations for acquiring a transcendent fix via hierophany is the Maya Pyramid of Kukulcan (the feathered-serpent god), also known as El Castillo, at Chichen Itza in Yucatan. Chichen Itza is fairly easy to get to—just a 2.5-hour ride from Cancun on a superhighway and only 1.5 hours from Yucatan's capital city of Merida. If you are there on the afternoon of the spring equinox, you can witness, as I have, the shadow of the "descending serpent" cast on the northern balustrade by the northwest corner of the stepped pyramid (Figure 1). A sculpted serpent head at the stairway's base adds to the drama of the imagery. Every year on March 20 crowds numbering in the tens of thousands fill the vast plaza surrounding the pyramid to witness the spectacle, today presided over by government officials. There are dancers, musicians, groups of meditators, and hosts of sacro-tourists, many of them North Americans and Europeans.

I first started following the way of the serpent of light shortly after coming across an obscure 1970 note by an obscure figure in an obscure journal printed in mimeograph that reported the phenomenon.[2] At first it was the astronomy that interested me. How did the hierophany work? Was it planned? Did the Maya make it happen or is the descending serpent just the product of the overworked imagination of some contemporary traveler? Then my interest shifted to the *hierophants*, the people who go to Chichen Itza to witness and participate in the spectacle.

I have done the hierophany—bought the t-shirt as they say—at least a half dozen times since then. Let me share my recollection

1. Sacred travel abounds as Y12 approaches. You need to be in the right place at the right time to get the Maya transcendent message. Here thousands of tourists assemble at the Maya ruins of Chichen Itza, Yucatan, on the afternoon of the spring equinox to watch the serpent descend. The image of the ancient Maya deity appears as a light-and-shadow hierophany on the northwest balustrade (*left*). Note the open-mouthed stone serpent head at the base of the half-diamond-shaped images. (Photo by George Keene)

of one such occasion. It was four o'clock on the afternoon of March 20, 2001, the first equinox of a new millennium. I stood there with 45,000 others in the plaza. We came by car, bus, train, plane, and cruise ship from all over Mexico, the Americas, Europe, and around the world—religious people, scientific photo-documenters, vacationing tourists, people in groups, families, solitary people. White people and black people come to Chichen Itza; mestizo people and people with goods to sell and ideas to trade stood alongside others looking for guidance, direction, or just a good time.

We all came to Chichen Itza to watch the magical interplay of the zigzag of light and shadow cast by one edge of the stepped seventy-five-foot-tall pyramid on itself. At four o'clock the shadow was just about to take on the shape of a giant snake. A number of those I had talked to who had seen it before told me that when they glimpse the luminous serpent made up of seven half-diamond-shaped patches of light, they share a moment in time with the ancient Maya, for legend has it that the ancients too witnessed that same image a thousand years ago alighting upon this most monumental of all their sacred works. Kukulcan was the Maya god of rejuvenation and his effigy symbolizes the renewal of life.

There we all waited, poised behind a chain fifty feet from the pyramid; it had been put there to prevent eager onlookers from ascending the steep steps and breaking the mood of anticipation that descends on the onlookers as the image gradually morphs into its serpentine shape. Just a few minutes past four the first hint of a pattern made its appearance on the stairway. In the 90 percent humidity most shadow seekers had not really settled in, although pilgrims who had arrived in the early hours of dawn had been sitting for hours on mats, towels, or pieces of cardboard cartons on claimed turf. Since early afternoon, colorfully garbed native dancers from the Folkloric Ballet of the state of Yucatan had entertained the crowd. Everyone listened to the orchestra play authentic Maya music. We witnessed a dance aimed at drawing the serpent down from heaven so that his energy might rekindle the spirit of life within us. We thrilled to a theatrical performance on Maya prophecy and the failed nineteenth-century Yucatecan resistance movement against Spanish colonialism. Then came a showy speech by the governor of the state of Yucatan—all of it broadcast from a grandstand off to the southeast side of El Castillo. In the interim some had raced back to the shops at the entrance for a quick snack; others, fearful of losing their places, picnicked on their dusty, prized square meter while dodging the interweaving pedestrian traffic that flowed zigzaggedly, like the famous snake, about Kukulcan's temple.

Four thirty and the first few elongated diaphanous triangles that would ultimately make up the ophidian shape were fully formed at the top of the balustrade. The muffled voice of archaeologist Alfredo Barrera Rubio, then director of the regional center of the National Institute of Anthropology and History (INAH), began the official play-by-play audio account that always accompanies the annual appearance of this equinoctial ribbon of light. He timed the appearance of each of the lighted geometrical figures as they took shape one by one from top to bottom down the side of the stairway. By four fifty-five all seven half-diamonds of light stood in place, the last one seeming to attach itself to the large open-mouthed serpent's head carved in stone at the base of the temple.

The crowd sat transfixed. All movement and sound abated as the luminous triangles linked together and slowly begin to slide toward the upper edge of the stairway's balustrade as the sun plunged downward toward the horizon. Late afternoon shadows lengthened and the air began to cool just a bit. A puffy cumulus dimmed the sun for a few moments, but when dazzling sunlight returned, the first of many collective *oooohs!* greeted the appearance of an even sharper image of the feathered serpent deity. A half dozen elderly New Agers collected together near the restraining chain and chanted in unison as the sound of a beckoning conch trumpet momentarily broke the silence. Bare-chested, long-haired white men, eyes closed, raised their hands in the direction of the sinuous image. Guards drove back the one or two zealous fans of the serpent who could not resist jumping the chain in a futile attempt to lay hands on the façade they had deemed holy. A fair-haired woman with Scandinavian features held a naked one-year-old child over her head above the crowd and directed his countenance toward the pyramid, seeking to bathe him in serpentine energy, while intervals between camera-shutter clicks waxed to an almost continuous low-pitched rattle.

Five fifteen and as the sun continued to dive earthward, Kukul-can's façade became all ashade but for the thinning, illuminated, undulating swatches. More *oooohs!* and *aaahs!*—the chants and mantras reached a five-minute-long crescendo as the serpent phenomenon began to fade from view. Well before the last luminous segment of slithering serpent slipped off the balustrade and vanished into the sky, some pyramid watchers began to head for the parking lot. By five thirty the show was officially over and what had been a trickle of exhilarated pilgrims exiting the ancient ruins, assured of the continuation of the cycle of life, now turned into a crush of tired tourists advancing like so much freeway traffic during the rush hour. Guides waved signs of variegated colors and symbols above the crowd to keep their tour groups together. More indefatigable sojourners piled into the gallery of shops outside the gate to collect their last souvenir t-shirt or ceramic idol, while dozens of bus drivers revved their diesel engines in anticipation of the hourlong, one-mile, traffic-clogged meander back to the main highway. The first equinox serpent hierophany of the new millennium was over. But like a rejuvenated snake having shed his skin, the serpent would return for the next rite of spring that follows the completion of time's annual cycle. The whole springtime Maya spectacle—serpent *and* people—was a wonder to behold.

Mircea Eliade, the historian of religion who invented the term *hierophany*, once wrote that spring is a resurrection of all life. In that cosmic act, all the forces of creation return to their first vigor. Life is wholly reconstituted; everything begins afresh; in short, the primeval act of the creation of the cosmos is repeated, for every regeneration is a new birth, a return to that mythical moment when for the first time a form appeared that was destined to be constantly regenerated.

Fueled by the Chichen Itza hierophany, archaeological sites all over Mesoamerica now draw huge crowds on the spring equinox. Yucatecans who live in the capital go to nearby Dzibilchaltun to see the equinox sunrise through the doorway of the House of the Seven Dolls, whereas those in northern Mexico assemble at Alta Vista, near Durango, where the Sun Temple's corners and a labyrinthine walkway line up with the rising sun. Enthusiasts from the heavily populated Mexico City area assemble at the great pyramids of Teotihuacan, although nothing celestially noteworthy actually transpires there on March 20.

I told my personal story of witnessing Chichen Itza hierophany to make it clear that even a rational-minded individual such as I can be moved by the shared experience of being in a place deemed sacred—especially at a moment when one of time's cycles begins anew. But why should Maya's Chichen Itza or the Druid's Stonehenge or the Inca's Machu Picchu—all sacred places to different cultures of the ancient world—be sacred to *us*? I think Gnosticism, a form of religious internationalism (which I will explore in greater detail in Chapter 6) may be a unifying force.

Long a part of Western history, Gnosticism embraces the idea that higher knowledge, not faith, is the key to salvation and that it can be acquired by ordinary mortals. Between the first and sixth centuries AD, for example, Gnosticism served as a middle ground between paganism and newly emergent Christianity. The Gnostics were a class of sects who believed, unlike their mainstream Christian counterparts, that it was ignorance rather than sin that cut us off from union with the creator. They devoted themselves to what they termed the "search for true knowledge." Instantly revealed truth, they said, could be found among *all* civilizations and every faith contained a germ of truth that culminated in Christ. But the early church fathers regarded Gnosticism as dangerous, particularly because of its adherence to magical practices, such as using talismans, secret phrases, and codes thought to be embedded in sacred texts (*The DaVinci Code* and its sequel come to mind).

Gnostics reasoned that God, who created everything, was also responsible for the evil in the world. There is a whole world of spirits between Him and us and it is out of *their* sinning—not ours—that the world had arrived at its corrupt condition. But we can take action and seek salvation through the psychic or adept, the one with the latent capacity to sense true knowledge, the one with the magical passwords needed to ascend the ladder of the demon-filled planetary spheres toward heaven and redemption. One scholar has characterized Gnosticism as a mixture of Eastern religions couched in the language of Greek philosophy and originating in "an atmosphere of intense other worldliness and imaginative myth making."[3] These words fit today's 2012 wisdom seekers like a glove.

Contemporary Gnosticism straddles the boundary between science and religion. Often laced with scientific language, this new brand of Gnosticism is built around the basic idea that *all* existences originate in a higher power that manifests itself by successive emotions that take the form of temporal turning points, or turnovers, of eons. Ecstasy or catastrophe? The prognosis for these stressful times is usually doom, hopefully accompanied by salvation or self-realization, provided we do the right thing, be in the right place at the right time—*connect*. Let us look at some of these professed, latter-day wisdom seekers and examine the sources of their insights so that we can get a better handle on the proposed meaning of 2012.

Geoff Stray is Y12's encyclopedist. He bills his vast, ominous-sounding Web site 2012: Dire Gnosis as the fast track to understanding 2012. As advertised, it "takes the most significant and fascinating parts of the 2012 puzzle and packs them into a bite-size package that today's busy people can digest in their lunch breaks."[4] Products for sale on his Web site include 2012 t-shirts, Frisbees, mouse pads, wall clocks, office mugs, underpants, and thongs. The site also includes his video "Beyond 2012—Game Over or Next Level?"

If Stray is the compiler of Y12 gnostic mythology, the brothers McKenna, Terrence and Dennis, surely were among its founding fathers. In their 1971 booklet *The Invisible Landscape: Mind, Hallucinogen, and the I Ching*, they tell of an Amazonian odyssey during which they tripped on sacred plants related to natural secretions of the pineal gland especially pronounced in advanced meditators—like Buddhist monks. Thus they acquired insight into the *I Ching* as "a mathematically coded form of the time wave system that underlies change in the universe."[5] The McKennas believed the *I Ching*'s sixty-four hexagrams corresponded to the sixty-four "codons" in human DNA.[6] They also lay claim to the discovery of a complex fractal wave that works in multiples of sixty-four, which leads to a series of levels that describes *all* changes in the universe, from subatomic all the way up to universal time, the sub-waves peaking together in 2012. All of this, mind you, is independent of any knowledge about the Maya calendar.

The movement toward contemporary Maya-based Armageddon exploded full force in a 1987 book titled *The Mayan Factor* by art historian Jose Argüelles, although he had his predecessors a decade earlier, as we will discover in Chapter 7. Argüelles claimed that Western technology and the corrupt civilization that gave birth to it are doomed—unless we shift gears at one of the seminal moments of synchronization of human and cosmic time. Argüelles plows through a series of complex calculations and charts that lead to the conclusion that the first entry pointed to a great "Harmonic Convergence" that was to take place on August 16–17, 1987. The date Argüelles landed on, although based on manipulations of the Aztec calendar, is supposed to be a prelude to what awaits us at "beam's end" in 2012. The galactic beam Argüelles refers to is the one the Maya (who, he argues, are actually descended from extragalactic beings) habitually latch onto as a means of conveyance around the universe from planet to planet, sowing the seeds of civilization as they go. On convergence day 1987, one of Argüelles's calculated Maya temporal nodes, the aliens were due to return to tend their crops. He suggests this would be the most opportune time for us to reconnect with the "heliotropic octaves" in the "solar-activated electromagnetic field," which will "cause the senses to attain new revelations."[7] Above all, says Argüelles, the defining moment will liberate us from the negative influence of Western science and technology.

On Sunday morning, August 16, 1987, thousands of Argüelles's followers held hands on Mount Shasta in the state of Washington, on Mount Fuji in Japan, on Machu Picchu in Peru, and at Egypt's pyramid of Khufu—all acupuncture points, or "all planetary light-body grid points in the body of mother earth"—to prepare for the mass cultural reawakening, "a reimpregnation of the planetary field with the archetypal, harmonic experiences of the planetary whole."[8]

In a promised instant, Argüelles predicted, all fear will turn to love, all separation to unity. The evil Mexican god Tezcatlipoca will reveal himself to be Quetzalcoatl,[9] god of peace and love, and the "New Age" shall dawn upon us.

There are no records regarding the appearance of UFOs at any of the sites of the prophesied hierophany, although a few of Argüelles's disciples claim to have experienced their transforming quality. Also, a number of celebrants said they had dreamed that night of energy flowing into their bodies as well as those of the children of the new generation.

As I said, 1987 was just a tune-up for what is in store for us 2012. Since then, prophet Argüelles, who now goes by the name of Valam Votan, Closer of the Cycle, has devised his own "Dreamspell" calendar, which he advises all cultures of the world to adopt. Votan's calendar comes replete with daily predictions based on the user's date of birth. Argüelles's personal astrological predictions are distinctly linked with four-dimensional galactic time travel. He envisions a world of telepathic communication emerging out of Y12—no talking, no writing, not even a technology (beyond solar). Anybody who does not evolve socially and spiritually will not survive in this peaceable kingdom. Instead they will be hauled away in silver space vehicles to a higher life on a distant planet called home (which beats burning in hell!). Although I have spent years studying Mesoamerican calendars, I must confess that I cannot understand even one of Argüelles's complicated-looking diagrams. Nor can I follow his explanations, which, like the McKenna brothers' and so many other 2012 narratives, is punctuated with scientific jargon incomprehensible even to scientists.

A more down-to-earth scientific-sounding prognosis, although with mystical overtones, for timing the advent of earthly paradise emanates from the articulate pen of John Major Jenkins, a software engineer. In *Maya Cosmogenesis 2012* and other works, Jenkins claims to have acquired his insight into the magic Maya moment in our shared immediate future when he was in his early twenties sitting atop one of the Maya temples at Tikal in Guatemala. He tells us that he heard "the wind whisper messages of a far off time and of another world."[10] It was the first time he had visited an ancient Maya site.

Fascinated by Maya calendrics, Jenkins began to seek an astronomical basis for setting time zero in the Maya calendar. He came to the conclusion that the Maya had geared their grandest timekeeping cycle to a starting date connected to an astronomical convergence that they predicted would happen some 5,126 years later—the alignment of the December solstice sun (the southernmost annual standstill point of the sun) with the heart of our Milky Way Galaxy, which is located in the constellation of Sagittarius.[11] In order to be able to accomplish such a feat, writes Jenkins, the Maya would need to have known about the 26,000-year period of the precession of the equinoxes, the slow movement of the direction of the earth's axis of rotation among the stars. The Maya Long Count happens to be equivalent to approximately one-fifth of this cycle. Jenkins backs up his hypothesis with impressive looking graphs and geometry. To bolster his galactic Long Count calendar theory he cites evidence in the ancient Maya record, including sculpture and inscriptions as well as alignments in Maya architecture.

Jenkins finds clues to the re-creation expected in Y12 pictured on Stela 25, located at the ruins of Izapa,[12] near the Pacific Coast of southern Mexico. Stela 25 (Figure 2a) pictures a humanoid figure supporting a staff with a bird perched on top. At the base of the staff lies the head of an alligator (or caiman) bound to a tree; his body extends upward, parallel to the staff. The bird in the tree has been interpreted by some Mayanists as the Big Dipper. The head of the caiman, Jenkins argues, is the "head of the Milky Way," or its center in our constellation of Sagittarius, where the nuclear bulge of the Milky Way is situated.[13]

The lineup along the local north-south meridian of the great cosmic tree that connects bird and beast is allegedly mapped out on the sky as it appeared at local midnight on the summer solstice at the time Stela 25 was erected (ca. 300 BC) (Figure 2b). Jenkins believes that the scene both on the stela and in the sky represents a

a

c

**2.** Is the Maya end of the world mapped on the sky? *a.* Stela 25 at the ancient city of Izapa may show a scene from the *Popol Vuh*, a book about the Maya creation, in which one of the hero twins confronts the Bird Deity, Seven Macaw, who has pretenses of becoming the new sun, shortly before the present creation (courtesy of the New World Archaeological Foundation). *b.* A modern software rendition of the sky shows the Milky Way aligning north-south. Does this map the alligator-tree scene pictured on Stela 25? (D. Freidel, L. Schele, and J. Parker, "Maya Cosmos: Three Thousand Years on the Shaman's Path" [New York: William Morrow, 1993], 77; ©1993 David Freidel, Linda Schele, and Joy Parker). *c.* The same creation scene has been likened to the imagery on the Blowgunner Pot, which shows the bird perched in a tree, just about to be zapped by a hero twin's shot from a blowgun. Off to one side lies Scorpius as represented (backward) on the sky map. The Big Dipper plays the role of Seven Macaw (© Justin Kerr, mayavase.com, file no. K1226).

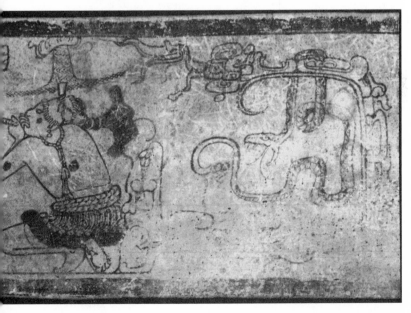

seminal Maya cosmic dialectic (despite the fact that Izapa is not a Maya site) that occasioned a great cosmological shift away from a system centered on the polar region to one focused on the southern Milky Way. He bases much of his case on Izapa being the place where both the galactic-solstice alignment was first celebrated and the Long Count originated. Jenkins cites the alignment of the Izapa Group F ballcourt as further proof. (I will discuss this alignment in detail in Chapter 3 and the theory of cosmic shifts and star-fixed world ages in general in Chapter 6.) Izapa monuments fronting the Group F ballcourt allegedly provide further clues to a 2012-timed creation; for example, there is a solar deity paddling down the Milky Way in a canoe and a snake-mouthed ballcourt marker. Jenkins wonders, can it be just a coincidence that the people who built Izapa had focused so intently on the horizon where the winter solstice sunrise lines up with the Milky Way?

The astronomical positioning is explained in the *Popol Vuh*, the sacred book of the Quiché Maya creation, a story that appears in the post-conquest literature and tells of a battle between the hero twins—the sun and moon of the present creation—and a pretender solar deity in the form of a bird named Seven Macaw.[14] He is the Big Dipper bird in the cosmic tree as well as on the so-called Blowgunner Pot in Figure 2c, which Jenkins considers the third medium (in addition to sky and stela) that tells the tale of creation. The twins shot the boastful faker out of his perch in the tree with their blowguns. Later in the story, the twins defeat the Lords of the Underworld in a ballgame, which served both as an elite enterprise heavily laden with ritual as well as a popular sport. The twins win the day and dissolve into the sky where they become the sun and the moon (or Venus). Most Maya scholars agree that these and other feats of cleverness and prowess emerge as the qualities of the heroic deities responsible for the present creation, qualities meant to be imitated by the heads of the chosen lineage the gods created to care for "all the sky earth," as they referred to their universe.

I think Jenkins's theory, which seems persuasive enough at the surface, has attracted a lot of popular attention because of the

Gnostic overtones that resonate in his explanation of what his alignment event means for us. He tells his readers that when the time arrives, his alignment will offer us an opening in time—a sudden new revelation about what it means to be alive. "What will emerge is persuasive evidence of the Maya's ancient and profound understanding of cosmological processes, including galactic forces that impinge upon the evolution of life on earth.[15]

Jenkins interprets the Milky Way as the mother of creation; the dark rift or apparent opening in the luminous bulge near the galactic center is her womb because "[i]t makes us think of an engorged, womblike area, easily giving rise to the idea that the Milky Way is a huge, pregnant being, and the central bulge is then the womb or birthplace of the sky."[16]

Jenkins theorizes further about what will happen at the time of the great crossover. It might produce a field reversal, not unlike the reversal of the earth's magnetic field. Whatever the physical outcome, the effect of the moment of alignment will mark an entry into a new world age that will transform our state of consciousness into a kind of collective earth spirit—provided we are ready for it. He adds: "The more who make the journey, the bigger the spirit-magnet gets, until we have all been drawn back into the cosmic heart. Returning to our daily lives renewed and realigned with the Creation Place, we'll bring the Galactic wisdom and a little bit of eternity down to earth."[17]

Like Argüelles's harmonic wave theory, the idea seems to be that celestial harmony is maximized when the sun at the solstices and equinoxes aligns with the Galaxy. Jenkins never reveals the mechanism that will make it all happen (to judge from his language, it seems to have something to do with magnetism), but he is dead sure that the last time the autumn equinox fell out of alignment (some 6,400 years ago) all of humanity responded by descending into warfare and conflict. That Jenkins's ideas have not been well received among mainstream Maya scholars, who place little stock in subjective analogies and knowledge acquired through revelation, comes as no surprise. Meanwhile, freelancer Jenkins responds by disparaging the academic community of Mayanists who, he says, have shut him out and ignored him.[18]

Other twenty-first-century peripheral prophets claim to possess secret knowledge of the ancient Maya cosmos and its possible effect on our future, but many of them (unlike Jenkins, to his credit) pay little or no attention to the remains left behind by ancient Maya culture. Wikipedia sports an entry that covers the entire disorganized panoply of New Age beliefs about Maya people: "Mayanism."[19] (Unfortunately, this term is often confused with Maya studies and Mayanists, who study the Maya culture.) Focusing initially on connections between the ancient Maya and secret knowledge of lost civilizations, such as Atlantis, nineteenth-century Mayanism has evolved into its contemporary form often by alluding to possible contact with extraterrestrials as well as to aspects of intelligent design. Among the enlightened latter-day proponents of Mayanism's beliefs is Carl-Johan Calleman. Calleman describes himself as a biologist and a cancer researcher associated with the World Health Organization. He proclaims that the "harmonic coincidence" attending the seminal 2012 moment will constitute nothing less than a spiritual awakening, an enlightenment that demonstrates the progress of evolution.[20] He appeals less to the book of Darwin and more to the book of Daniel and other Biblical prophets. Like his predecessor Argüelles, Calleman advocates meditating in large groups and mass hand-holding and hugging when the time comes, as a way to break through to the other side (to borrow rocker Jim Morrison's musical epithet).

For Calleman, the Maya step in time is all part of a vast mathematical progression grounded in social Darwinism and the theory of social progress. The nine levels typical of many Maya pyramids, like their nine underworld levels, represent the hierarchical structure of time dating all the way back to the Big Bang, which he pegs at 16.4 billion years ago. Steps up the pyramid connote ever-rapidly evolving states of consciousness; thus, step two marks the evolution of animal life (820 million years ago); step three, primates (41 million); step four, *Homo sapiens*' tribal organization (two million);

step five, spoken language (102,000); step six, the creation of a patriarchal civilization with laws and a written language (5,125); and step seven, industrial technology (256), which brings us to AD 1750 via seven levels of progress. (Note the progression in time divisions by twenties, the base of the Maya counting system.) The eighth level, which began in 1999, was marked by the introduction of the Internet and the infrastructure of global communication. Calleman offers us his own calculated termination date, claiming there is an error in the Maya calendar. According to him the ultimate step will take place not on December 21, 2012, but rather on October 21, 2011. On that day we cross over into the Universal Underworld of Consciousness. That is when we take on the awesome responsibility of co-creator (with God). This is a task well worth preparing for, and links on Calleman's Web site offer hints on how to proceed: you should cleanse your colon with bentonite clay elixir so that you will be fully receptive to the "Pranic energy," or life force, that will arrive then to reset your body in balance.[21]

Back in the 1960s, when I began my career in teaching, I recall engaging in many intense discussions with students about the use of psychoactive drugs. We all agreed that LSD certainly activates your psyche—but to what end? The students split sharply into two camps drawn strictly along the line between those who had ingested the substance and those who had abstained. The former group claimed that drugs can transport you to a more elevated state of consciousness (whence the term "high"), a condition in which deeper insights are not only attainable but also creatively expressible—artistically, musically, poetically. One of my young charges was convinced that the tiniest bit of lysergic acid had once transported his brain all the way to the planet Saturn and back! The latter camp, comprising non-users along with a few failed experimenters, agreed with most neurochemists on how you attain alternative states. You hallucinate on hallucinogens; that means you experience false realities. If your sensate potential is altered, it is because of brain chemistry, the effects of chemicals on the neural system. Near-death experiences are like dream states; they are nothing more than images in a once-ordered brain, in this case messed up by the ingestion of mind-altering substances. So say the nonbelievers.[22]

Daniel Pinchbeck belongs to the first group. He believes in "LSD synchronized with the activation of the noosphere" (a sort of harmonized collective consciousness).[23] Writer Pinchbeck's take on 2012 is decidedly more metaphysical than Jenkins's and Argüelles's theories. In his book *2012: The Return of Quetzalcoatl* and on his Web site www.realitysandwich.com, Pinchbeck accepts Jenkins's galactic alignment, and then draws on his own interest in psychedelic shamanism (he had already produced an earlier book titled *Breaking Open the Head: A Psychedelic Journey into the Heart of Contemporary Shamanism*). Pinchbeck seems particularly attracted to the possibility, also mentioned by Jenkins, that ancient Maya acolytes trained in the priesthood of the new calendar sought and attained deeper insights through the use of psychedelic substances, such as the hallucinogenic psilocybin mushroom and psychoactive secretions of the toad *Bufo marinus*.

Attracted by the shamanistic aspects of the quest for pathways to the noosphere, Pinchbeck undertook personal journeys to faraway places, such as Mexico, the Amazon, and Gabon, in the company of native herbalists and healers. He experimented:

> On the night that Mars reached its closest approach [in 2005 Mars made its closest approach in some 6,000 years], I took a fungal sacrament with an old friend of mine from New York, a writer and theorist on potential uses of the internet to create "augmented social networks," linking progressive causes and affinity groups. . . . Hours after the trip ended I still found myself significantly altered; closing my eyes I beheld a monstrous entity—a Lovecraftian caterpillar creature with multiple heads and mutable human faces.[24]

Pinchbeck describes other materializations that connected him personally with a cosmic transmission signaling that the return of Mars "was a phase-shift, part of the process through which our planet was becoming, by subtly intensifying degrees, less materially dense and more psychically responsive." He states, "This was my transmission."[25]

Clearly, Pinchbeck portrays himself as an adept, one who believes himself capable of acquiring higher truths to which we ordinary mortals are simply not in tune. Like my students in the LSD debate or the ardent believer in astrological prediction who says, "All I know is that it works for me," you either accept what he has to say at his word or you reject it—no discussion.

———————

The feeling that there is "something in the air"—a sense that the world is due for big changes—emerges as a common theme in all the Y12 scenarios I have been reviewing in this chapter. Either we work at collective self-improvement to reclaim the psychic dimensions of ourselves or the planet descends into environmental ruin. So write the prophets.

We have had plenty of warnings. In his book *Apocalypse 2012*, Lawrence Joseph, a journalist and chairman of the Aerospace Consulting Corporation, gets my vote for ringing the largest number of Y12 catastrophic bells. This book will "make you think twice about your retirement plans,"[26] announces the provocative book jacket. There is a black hole at the center of the Galaxy, claims Joseph, and it is sucking up matter, energy, and time. (The intense contemporary preoccupation with 2012 galactic phenomena, which I will discuss in Chapter 7, is truly fascinating.) That means whatever energy arrives at our planet from the black hole will be disrupted for the first time in 26,000 years on December 21, 2012—at 11:11 PM universal time. According to Joseph, the Maya say the mechanisms of our body and our world will be thrown out of kilter.

Just look at all the coincidences: On October 1, 2006, Hurricane Stan made landfall and wreaked havoc in Central America. Seven days later a 5.8 magnitude earthquake hit the same area—all of that on the heels of Hurricanes Rita and Wilma. Unrelated events, asks Joseph, or signs of a bigger catastrophe to come? September 9,

2005: supergiant flares erupted on the surface of the sun—the latest in a series of spectacular solar events for that year. An extraordinary abundance of atomic particles was propelled earthward—and this, mind you, in a *minimum* year in the eleven-year solar cycle. The next maximum, by the way, is due in 2012. Then on September 14 there was an earthquake in Ethiopia. Could these multiple happenings be triggered by planetary alignments that periodically distort the shape of the sun? Joseph wonders.

To make matters worse, the next peak of the cycle will coincide with a great planetary tidal force—caused by a planetary lineup—Joseph adds. The prognosis is not good: these forces will unleash megabursts of "imprisoned radiation trapped inside the sun for thousands of years quite possibly far deadlier than any the earth has encountered since [the species] *Homo sapiens* has been around."[27]

The bad news from prophet Joseph does not end there: the earth's magnetic field is weakening and is about to undergo a reversal. Of course, it is a slow process, but eventually birds and fish will start to get lost during their annual migrations—and the tangling of magnetic field lines will alter the directions of hurricanes and tornadoes. But forget about the magnetic flip for a moment; like the holes in the ozone layer that protects us from harmful ultraviolet radiation, cracks seem to be developing in the earth's magnetic field, weakening our defense against cosmic radiation.

Like many Y12 doomsayers, Joseph's pitch is that of an ardent activist against establishment science, one who believes that both neglect and conspiracy permeate that community. For example, he labels archaeologists "cultural imperialists" for the gross inaccuracies they propagate about the ancient cultures whose remains they desecrate and analyze and for their failure to acknowledge the information about the past that can be acquired from native wisdom. The solar- and geophysicists receive a tongue-lashing too: why do they refuse to get together and address once and for all the obvious connection between sunspots and violent weather events? Do not be surprised, he warns, if there is a protest at the forthcoming International Heliophysical Year (2007) meetings—"a populist demand for a more complete disclosure of solar activity data—data vital to our personal and ecological health, which has been gathered exclusively with public funds."[28]

Compare Joseph's concerns with the words of John Major Jenkins: "Western or Euro-American civilization currently rules the globe through dominator forms of coercion and resource control. . . . Clearly our distant ancestors participated in a style of culture that is fundamentally antithetical to our own."[29] Or with Jose Argüelles's take on establishment science: "Entrenched and ever-vigilant in their self-support, the forces of scientific materialism have suddenly guarded the portals to their domain, keeping in mind a singular goal: to maintain the myth of ever-progressing technological superiority."[30]

Although 2012-ologists are caustically critical of contemporary science and technology at so many levels, I truly believe they are well-intentioned people who want to see the world change. Generally aligned politically left (they tend to be pro-environment, anti-corporate, and anti–consumer capitalism), most Y12 prophets come across to me as activists with a deep desire to participate in the process of evolution. They want to run their own cosmology. Paradoxically, even though they are anti-technology, they proliferate on the Internet: the last time I googled "Maya Creation 2012" (on June 3, 2009) I got 2,060,000 hits.

Suspicious of establishment science, the prophets of 2012 nonetheless make frequent use (and, more often, misuse) of scientific concepts. For example, we have already sampled the scientific lingo that peppers so many of their ideas. Argüelles's beams from the center of the Galaxy accelerate prophecies that need to come to fruition. Gregg Braden, another Y12 seer, thinks the earth's presently weakened magnetic field, which makes us more susceptible to solar flares, is the glue of our consciousness.[31] And when it is weak, we are open to change.

More broadly categorized, Y12-ologists are also anti-mechanization, anti-media, even anti-urban, and they are highly critical of establishment academia (or establishment anything for that matter).

Unfortunately, as I will show in Chapters 5–7, they are also *un*critical of their own ideas.

Most advocates of a big Y12 happening also coalesce around the notion that ancient galactic wisdom has been around across *all* ancient cultures (which is why I label it "Gnostic thinking") for a long time. The idea seems to be that all ancient cultures were aware that a great transformation of the ages would occur sometime in the future. The question is, Will we listen? And will we tune in? If we do not change our ways, Braden tells us, we will have a difficult time surviving in the new world that awaits us, so be nice or get mulched—or get carted away on a silver spacecraft!

Y12 hype has already had a profound effect on American pop-culture media. Numerous television documentaries address the question of what will happen on the December solstice that year. As 2012 approaches, a rash of end-of-the-world films floods U.S. theaters and video screens. Most of them focus on disaster and redemption. For example, released late in 2009, the film simply titled *2012* draws on the popular theme of parallel universes. It includes vivid special-effects apocalyptic scenes. Another film, *2012: The War for Souls* (forthcoming), is based on a novel by Whitley Streiber, a UFO contact proponent. In his version of doomsday, aliens come to earth to snatch souls but they keep the bodies alive for the purpose of slavery. Doubted by everyone, the film's hero single-handedly seeks to break through to a parallel earth where he can stop the invasion. The end of the *X-Files* TV series extols the world-changing Y12 event as the colonization of earth by aliens. Many contemporary films replay the scenarios laid out in their earlier counterparts. In *The Day after Tomorrow* (2004), for example, the impact of technology on the environment leads to global disaster. It grossed more than half a billion dollars at the box office. In the award-winning animated sci-fi film *WALL-E* (2008), a humanistic robot of the future roams the trash heap of a future polluted world in search of meaning. A female robot, aptly named Eve, helps him find it. *Indiana Jones and the Kingdom of the Crystal Skull* (2008) and *I Am Legend* (2007) also proffer a sudden end of the world unless particular human action is taken (in these two cases, collecting crystal skulls and dealing with disease-induced vampires, respectively). Finally, in the prescient *Death Race* (2008), a failed economy is the doomsday culprit.

Survival kits are now available for Y12, as well as items of proper attire for the moment, including a multitude of t-shirts bearing slogans such as "Doomsday 2012" and "Shift Happens." There are books too, like Christy Raedeke's *Prophecy of Days*, a *DaVinci Code* / Maya calendar fusion. On the nonfiction side there is *The Complete Idiot's Guide to 2012*. Plans are underway for sacred travelers to be in the right place at the right time, á la the descending serpent at Chichen Itza. For example, leading up to the Third International Copan Star Party, slated for December 21, 2012, will be special events on the winter solstices in 2010 and 2011 to gear up for the big one at the famous Maya ruins in Honduras. Proceeds go to a worthy cause: building a science center for kids in a nearby town. I have been invited to attend a conference tour in Tikal, Guatemala, on the seminal date. In 2007, a 2012 New Age conference in Hollywood drew 1,000 participants. Another in San Francisco focused on the idea that we are entering the greatest crisis in all of human history.

Although the Maya seemed to exhibit little knowledge of their ancient Long Count calendar when queried in the nineteenth century by the first outsiders to visit Yucatan, some of their modern descendants offer what they claim to be authentic words of wisdom on the subject of 2012. Lawrence Joseph, for example, claims to draw his Y12 native knowledge from a pair of Guatemalan natives, Gerardo Canek Barrios and Mercedes Barrios Longfellow. They say they sought out Maya "elders," authentic card-carrying shamans who, they claim, have retained an intimate knowledge of the Maya calendar since its foundation. Gerardo, who calls himself an elder, does not view 2012 as particularly destructive. "We see it as the birth of a new system."[32] The story goes like this: our ancestors come back; they are reincarnated every time someone is born. By 2012 they will all have returned and the cycle of death and rebirth will be complete. There will be tests—in our understanding, in harmony. Drawing on what sounds like borrowed Tibetan Buddhism, the Barrios' claim that we will need to share both our pain and our happiness. If we meet the tests, if we possess the wisdom, then we progress to a more enlightened era. If we do not, we will need to wait 5,125 years for another shot, for that is when the doorway to the next world age opens up. Says one elder:

> The world is transformed and we enter a period of understanding and harmonious coexistence where there is social justice and equality for all. It is a new way of life. With a new social order there comes a time of freedom where we can move like the clouds, without limitations, without borders. We will travel like the birds, without the need for passports. We will travel like the rivers, all heading towards the same point . . . the same objective. The Mayan prophecies are announcing a time of change. The Pop Wuj, the book of the Counsel, tells us, "It is time for dawn; let the dawn come, for the task to be finished."[33]

Adds another Maya seer:

> If humans don't correct our course in the face of these events we will be off-balance in the moment the event appears, a very strong event in comparison with what we have experienced. Humans more than ever should pay close attention to all the events that disturb balance. They are teachings that we living beings should extract from the stages through which we pass.[34]

Although I remain skeptical of these native prognoses, which seem designed to inspire sacred travelers, nonetheless they do seem less tinged with fear than most of the Anglo-based versions we sampled earlier. Although both native and Anglo versions share that sense of anxiety about impending change, in the native case that change seems more directed toward communal awareness and advocating human action and participation as key to restoring balance in the order of things. As we will see in the next chapter, this idea of balancing the cosmos actually squares pretty well with what scholars have learned from the study of Maya documents. As

I reread these quotes, I think I can understand why outsiders might find them appealing, particularly if they reflect their own beliefs that perceived injustices, secret conspiracies, and general evil exist in our decaying contemporary culture.

The contemporary receptivity to Maya wisdom makes it necessary for us to learn a bit more about ancient Maya culture and the calendar—especially if so many people are willing to hold it responsible for whatever will happen on that winter day looming on the temporal horizon. Therefore, in the next two chapters I will explore what I think we *do* know about the Maya based on material evidence. Sadly our resources are few—only three or four original written documents, the codices, survive from the era before European contact. The Spaniards burned them all in the sixteenth century out of fear that the Maya had been resisting conversion to Christianity because they were devil worshipers. On the plus side, we have access to numerous (now largely deciphered) inscriptions carved on monuments that adorn the magnificent architecture of Maya sites such as Tikal (Guatemala), Copan (Honduras), Palenque (Mexico) and dozens of other ancient ruins, many of which still lie buried in the rain forest of Mexico and Central America. Finally, in addition to the aforementioned *Popol Vuh*, we have access to the valuable Books of Chilam Balam (*chilam*, from *chi'* meaning "mouth"; and *balam*, a term meaning "jaguar" used for the jaguar priest). These prophetic histories were written in final form in Yucatec Mayan using the Roman alphabet in the seventeenth century or later. Although affected by the Hispanic world view at that time, many of these books contain veiled accounts of ancient Maya customs. We will use these resources in the following two chapters to explore the culture of the Maya, especially their calendars and their ideas about creation.

# WHAT WE KNOW ABOUT THE MAYA
# AND THEIR IDEAS ABOUT CREATION

"And then when the destruction of the world was finished
They settled this [land] so that Kan Xib Yui puts it in order
Then the White Imix Tree stands in the North
And stood as the pillar of the sky
The sign of the destruction of the world . . ."[1]

The Maya thought a lot about the creation of the world, as this passage, one of many from the colonial *Book of Chilam Balam of Chumayel*, exemplifies. But who were they and where did they come from? The Maya lived—and still do—in the peninsula of Yucatan, which encompasses portions of Mexico, Honduras, and El Salvador and all of Guatemala and Belize. They inhabit the southern end of Mesoamerica, a common culture area that exhibits a long tradition of a shared set of symbols and ideas, as well as social customs and material forms of expression. This larger cultural area stretches roughly from the Tropic of Cancer, just south of the U.S. border, all the way to the middle of Central America. The more we learn about expressions of Maya legitimacy—their art, architecture, and calendars—the more direct ties we find with the other cultures that made up ancient Mesoamerica, such as the Olmec, the Zapotec, the Aztec, and the people of Teotihuacan. Our own studies of ancient Mesoamerican calendars reveal specific attributes that cross over between the codices from the highlands of central Mexico and Maya Yucatan.[2]

Like all Native American people, the Maya descended from nomadic hunters following large game animals that crossed the Bering land bridge in successive waves of migration from Asia into Alaska 20,000 or more years ago. An alternative theory, which claims they came from the east across the north Atlantic, lacks sufficient proof, in my opinion. By the Early Formative period (2500–2000 BC) in Mesoamerica, people began the transition to agriculture with the production of maize and to the development of ceramic traditions. Around 1500 BC the Olmec center of San Lorenzo began its florescence, followed by La Venta, along the Gulf Coast west of the Yucatan peninsula.

In the Middle Formative period (900–400 BC), settlements sprang up in the Valley of Mexico and Oaxaca in the mountainous regions to the west of the Yucatan. We can document the first concrete achievements in calendar and astronomy from about this time. Carved stelae dated to the latter half of this period herald the beginnings of hieroglyphic writing, as well as use of the 365-day year and the unique 260-day cycle. (We will look at how these cycles operate in Chapter 4.) Great architecture and sculpture proliferated, accompanied by increasingly complex, hierarchically organized political and social systems—what cultural anthropologists call a *state*-level society. At about the same time a complex of cultures in the isthmian region of the Pacific Coast just south and southwest of Yucatan began to produce ceramic and architectural works. The city of Izapa rose to prominence as a regional capital on the periphery of this area during the late part of the Middle Preclassic period (300 BC–AD 200). All of these people and places contributed to the artistic styles; formats for laying out ceremonial centers with their impressive plazas, pyramids, and temples; and various early forms of pictograms and rudimentary hieroglyphs that coalesced in the ancient Maya culture, and these achievements have come to fascinate our contemporary culture.

As to astronomy, one of the earliest carved upright stones bearing information about a complex knowledge of the heavens comes from the Gulf Coast site of La Mojarra. Stela 1 contains hieroglyphs that highlight a pair of solar eclipses and first appearances of the planet Venus in its 584-day cycle; the monument dates to the second century AD. Periodic Venus sightings would have required at least a century or more of careful observation to yield a reliable pattern so that precise predictions could be made. So we know that sophisticated, precise calendar keepers were active at least as early as the beginning of the Christian era.

The pinnacle of complexity in the material works of the civilizations of Mesoamerica occurred during AD 200–900, a time when Europe slept in intellectual darkness. The Maya Classic, or Florescent, period was marked by the appearance of highly organized settlements, an advanced calendar, a complex religious pantheon, the rise of a social elite class, and government by dynastic rule. Nowhere were the qualities of civic and intellectual achievement more outstanding than in the area in and around the Petén rainforest of northern Guatemala. Tikal's skyscraper pyramids, Copan's exquisite high-relief sculpture, Palenque's delicate stucco hieroglyphs, and the detailed references to timekeeping on stelae, as well as the precise astronomical predictions in bark-paper codices, are works of great intellectual sophistication, unsurpassed in the Old World. Although the surviving codices are products of the Postclassic period (after AD 900), they refer to earlier dates on which astronomical calculations were based. Therefore, we can infer with some certainty that Maya books about timekeeping must have been around well back into the Early Classic period.

The core of Classic Maya society was built around a dynastic system reminiscent of the ancient Chinese and Egyptian civilizations, or that of the Hapsburgs, the Romanoffs, and the Windsors, familiar to us from European history. Hundreds of carved monuments erected throughout the Classic period prove that Maya royalty linked their ideas about rulership to an intimate knowledge of the workings of nature, including the heavens, which they believed certified their power and status. The Maya ancestors, the source of all sustenance for the common people and the root of the holy status of the revered rulers, could be accessed only through the afterworld, where they resided. The blood of kings, displayed through elaborate sacrificial ceremonies duly recorded in the monumental record, served as the lifeline of a contractual arrangement between ruler and transcendent ancestor deities in the sky.

Ritual offerings—*debt payments* as the Spanish chroniclers record them—were made in the awe-inspiring environment of the ceremonial center. The rites must have been quite impressive to the citizen onlooker. Imagine a Maya king seated on his inscribed throne in front of the doorway to his temple. Members of the royal lineage and his court surround him as he stoically deploys the spine of a stingray to pierce his genital member. His attendant collects the droplets of precious blood on parchment, and then burns them in a censer, from which the smoke rises up to the heavens. He has made his payment to the gods for the good fortune they have bestowed on his people. All the while, the masses of people assembled in the spacious plaza below witness the spectacle. What more can you sacrifice than your own blood? The written record and the study of temple alignments tell us that on occasion the ruler may have performed this penitential act as the very celestial body (i.e., Venus) from which he believed he drew his powers appeared in the sky over temple and throne—at the right time. The astronomer's calendar saw to it.

Calendar keepers were no workaday drones. They occupied prestigious positions in the ancient Maya court. The sixteenth-century chronicler of the Maya, Friar (later Bishop) Diego de Landa, tells us:

> The natives of Yucatan were as attentive to the matters of
> religion as to those of government and they had a high priest

whom they called Ah Kin [Daykeeper] Mai. . . . He was very much respected by the lords . . . and his sons or his nearest relatives succeeded him in office. In him was the key of their learning. . . . They provided priests for the towns when they were needed, examining them in the sciences . . . and they employed themselves in the duties of the temples and in teaching them their sciences as well as in writing books about them. . . . The sciences which they taught were the computation of the years, months and days, the festivals and ceremonies, the administration of the sacraments, the fateful days and seasons, their methods of devotion and their prophecies.[3]

There was grandeur in these rites that connected the world beyond with courtly events in the life of the ruler here below. Events commemorated included the initiation of a battle, a marriage, sealing a pact or peace treaty, the capture and punishment of the vanquished, a burial, and the fragile moment of transfer of power from a deceased leader to his offspring. For the Maya, life's seminal events needed to be timed and recorded precisely. One needed to seize the moment to achieve the balanced order of nature required in the harmonious governance of a stable society. The codices testify to it; they even depict items the people offered on momentous occasions—incense, turkeys, fish, jade, tamales, even blood (Figure 3). Each gift to the gods necessitated a specific offertory place and a proper time. The timing of the ritual—the correct day and season of the year—was always crucial.

Unlike our modern Big Bang cosmology, Maya creation was a participatory affair; that is, the Maya people believed they had a role to play in the outcome of things. Maya time was thought to have begun when the ancestor gods of the lineage that culminated in the current ruler subdued the lords of Xibalba, the underworld. Only then, as the world waited for the dawn, could the creation of man out of maize come about. As the *Popol Vuh* recounts, this was "the making, the modeling of our first mother-father, with yellow corn, white corn alone for the flesh, food alone for the human legs and arms, for our first fathers, the four human works."[4]

---

**3.** This scene from an almanac in the Maya Madrid Codex shows debt payments (headdresses or breastplates) being received by the gods. Pre-Hispanic sacred books connected with divinatory procedure, the codices are largely concerned with the appropriate timing of rituals during which such offerings were paid as debts to the gods to keep the world in harmony. (Akademische Druck-u. Verlag, Graz)

Today most of the ancient Maya cities lie in ruins, pretty much the way the earliest explorers found them shortly after Spanish contact in the mid-sixteenth century. Only a small percentage has been partially restored. As they enticed their discoverers, so too the Maya ruins invite us to romance them. What happened to

the Maya? Why did they fall? Those who speak of the celebrated "Maya collapse" often fail to note that Maya prosperity lasted a very long time. Under dynastic rule it persisted from at least the second century until the middle of the eighth century. The archaeological record suggests that the theory of a catastrophic collapse is vastly overgeneralized. In the first place, the institution of divine kingship was abandoned only in the southern lowlands. Secondly, many of the sites in that area actually manifested continuity, although their populations were reduced significantly. Amidst these changes, their methods of time reckoning, such as the Long Count, persisted.

The record of the inscription suggests that late in the ninth century the Maya lost their obsession for carving calendar dates on stelae, entombing their dead kings, and continually refurbishing their massive architectural works. But why did that happen? Archaeologists focusing on that problem believe crop failure caused by overpopulation and soil exhaustion, perhaps triggered by periods of drought, fueled the fires of social discontent.[5] Just as we are forced to give up luxuries in hard times, the Maya fell back to a simpler existence. There was a short-lived period of resurrection of the culture—less dynastically oriented—after the ninth century in northern Yucatan. By the time the Spanish invaders arrived early in the sixteenth century, there were only scattered native villages in general disunity. Still, large political units maintained centralized authority vested in elite patrilineages and alliances. Ironically, this discord prolonged and complicated the conquest of the Maya.

The myth of catastrophic Maya collapse has played a role in 2012 perspectives. For example, Mel Gibson's aptly titled *Apocalypto* (2006) incorrectly portrays the Maya as extremely violent, bloodthirsty people who lived in fear of reprisal by their terrifying gods, to whom they are compelled to offer vast quantities of blood from mass human sacrifice. The film also evokes the theme that the end of the world is imminent. The point of the story seems to be that here was a sophisticated civilization that achieved great heights in science, engineering, and mathematics (like us), but it also possessed a brutal and savage side (also like us). So *it* (read collapse/ Armageddon) can happen to us too.

Following the collapse, native people continued to live in and around the ruins.[6] They also continued to worship their shrines, and their rulers maintained the legends of their ancient past. The colossal architecture and carved monuments were subsumed by jungle, until adventurer-explorers from Europe and the United States rediscovered the lost Maya cities. Standing among the ancient ruins of Copan, nineteenth-century American explorer John Lloyd

Stephens could only speculate in wonderment at the fallen stelae, their impressive inscriptions carved in deep relief, and imagine a story of things that once transpired here, now lying hidden in the strange-looking script: "One thing I believe, that its history is graven on its monuments. No Champollion has yet brought to them the energies of the inquiring mind. Who shall read them?"[7]

Stephens correctly anticipated the outcome of a later age. The stelae would be read, but not until more than a century later, and the inscriptions, once deciphered, would tell not of men who came from Asia, Egypt, or Phoenicia, or of the Lost Tribes of Israel or the lost continents of Atlantis and Mu—some of the nineteenth-century fanciful explanations tied to the newly discovered ruins. Nor were the inscriptions the handiwork of placid Maya astrono-mer-priests who focused exclusively on deciphering the mysteries of the universe. In his tongue-in-cheek account of this era, archae-ologist Robert Wauchope, speaking of one post-Stephens theorist, comments, "He was utterly incapable of critically examining either the factual or the logical evidence bearing on any theory he wanted to believe."[8] The same might apply today to those who predict a 2012 apocalypse.

The glyphs are mostly about actual Maya history, or at least their version of it. The text of Copan's Stela B (Figure 4), for exam-ple, celebrates the end of a katun, which is sealed by the blood sac-rifice of Copan's most famous ruler:

> 9 baktuns 15 katuns 0 tuns 0 uinal 0 k'in 4 Ahau 13 Yax, [unde-ciphered titles such as God of Earth and Sky etc.] and then was erected the partition, the image of Macaw Mountain Lord.
> There were completed 15 katuns (when) he scattered drops (of blood) (in) the image of the sky God [ ], the 13th ruler of the dynasty, Waxaak Lahun Ubah Kawiil, Holy Lord of Copan.[9]

The other side of the stela shows the king, with double stingray spines for bloodletting holstered at his waist and the smoke from his incinerated sacrificial blood rising up to the mountain in the afterworld, where his ancestors reside.

In addition to the codices and inscriptions, books on divination and prophecy written in the early colonial period also provide a window into ancient Maya culture. For example, the Books of Chilam Balam from north Yucatan, a dozen texts written in the native Yucatecan language, are laden with prophesy, although they are also basically about history, as these passages suggest:

> Katun 8 Ahau [AD 687–692] Chichen Itza had been manifested . . .
>
> Katun 6 Ahau [AD 968–988] completed the seating of the lands of Champoton . . .
>
> Katun 8 Ahau [AD 1461–1481] They destroyed the governors of Chichen Itza.[10]

These books proclaim an abiding faith that what happened in any given katun is also destined to take place, in one form or another, in future katuns that carry the same number. To draw a parallel, we might say that the sixties are decades of assassination because Lincoln, two Kennedys, and King suffered that fate in those numbered bundles of time. The window to the past opened by the books of Chilam Balam, however, is slightly fogged, because these texts were heavily influenced by Christian concepts and Renaissance Spanish ideas brought over by the invaders. Still, scholars agree that many portions of the Books of Chilam Balam and other similar books are convincingly close to translations of parts of ancient codices.

---

**4.** *(Facing page)* Stela B, Copan, Honduras, shows the effigy of Waxaak Lahun Ubah Kawiil, the most famous Lord of Copan, garbed in royal dress and about to make a blood sacrifice with his stingray spines (arrows). Above him clouds envelop Macaw Mountain, the place of the ancestors. (See page 43 for a translation of the inscriptions on the reverse side.) In contrast to the hieroglyphs that appear in the codices, which are mostly concerned with divinatory procedure, monumental inscriptions carved in stone and placed in prominent locations at Maya sites are primarily about the exaltation of the ruler and his bloodline. (Drawing by Anne Dowd)

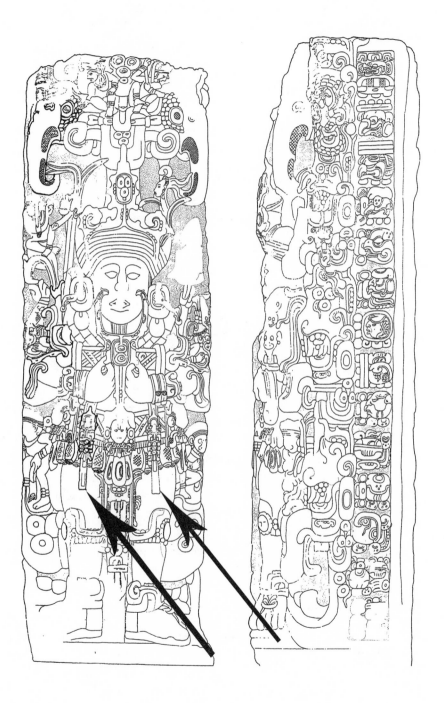

With this brief introduction to Maya culture and the resources to access it, the stage is now set for the all-important questions about creation, timekeeping, and calendar that bear on the Y12 phenomenon. Readers interested in further details about the ancient Maya should consult the references listed in my endnotes, especially the fascinating and informative works by anthropologist Michael Coe on how the Maya hieroglyphic code was cracked, art historian Mary Miller on the wonders of Maya art and architecture, and epigraphers Simon Martin and Nicolai Grube on the history of the Maya dynasties.[11]

What do we really know of Maya beliefs about the creation of the world? In tune with their cyclic calendar, the Maya believed that creations were cyclic and at least four in number. The only clear statement in the codices that might refer to the end of the world that I know of, occurs on page 74 of the Dresden Codex (Figure 5), named after the city where it surfaced in the nineteenth century. This page portrays a menacing scene showing water being vomited from the mouth of a sky serpent resembling an alligator or caiman, the animal that usually represents the celestial realm. More water emanates from the sun and moon hieroglyphs that segment the body of the serpent. These glyphic bands may represent the zodiac along which the sun, moon, and planets carry out their own cyclic journeys, or they may simply stand in for the sky in general. Still more water pours out of a vessel held by an old woman deity who appears suspended in the middle of the frame. Finally, at the bottom a male deity wields arrows and spears.

This scene may refer to the destruction of a previous world by flood, specifically the world that ended on August 11, 3114 BC,

---

**5.** (*Facing page*) The final page of the Dresden Codex, a pre-Columbian document from Yucatan, portrays the end of the last Maya creation in great flood. See page 48 for a description of the watery scenario. (Akademische Druck-u. Verlag, Graz)

and appropriately enough, it comes at the end of the document; but it might also signify a seasonal torrential downpour. Pursuing the destruction theme, the *Popol Vuh* tells us that the gods were dissatisfied with their creation of an earlier unsatisfactory human race—the wooden manikins they had fashioned to pray to them so that the universe would be kept on an even keel. The experiment did not work so they brought about a flood to destroy what they had created. The chocolate-brown background on Dresden 74 jibes with the *Popol Vuh*'s description of the resin that fell from the sky. Iconographers also identify the black spear-wielding god as a destroyer deity. The garbled inscriptions that accompany the picture are suggestive as well. They read: "Storm, black sky, black earth, first year . . ."[12] A similar flood event is described in one of the Chilam Balam books:

> And then great Itzam Cab Ain (the caiman sky deity) ascended
>     back then
> that this deluge may complete the word of the katun
>     (prophecy) series . . .
>
> One fetching of rain, (the flood being poured from a vase)
> One lancing of rain (a reference to the deity's spear)[13]

The Chilam Balam story then goes on to describe the next (fourth) creation, as I have quoted it at the opening of this chapter.

Now, is this "real history," in the Western sense of dated events, or is it metaphor? Does the narrative presage the actual end of the world? Or is it intended to serve as a framework, a template for the passing on of old ways about the purification and renewal that takes place at the turn of *all* time cycles, such as the New Fire ceremony that accompanies the completion of the calendar round (which I will discuss in the next chapter), or cycles of eclipses, Mars in retrograde, the appearance and disappearance of Venus, and so forth. Think of our New Year celebrations. We take account of ourselves by celebrating the end of our seasonal cycle—often with wretched excess—as the stroke of midnight approaches. Then we perform our penitential acts of cleansing or purifying ourselves the next morning, as we contemplate a brighter future, by making New Year's resolutions. Like those who have studied Biblical prophecy (see Chapter 6), scholars familiar with Maya philosophy are inclined to think that these prophecies were never intended to be interpreted literally.

Some Maya stelae carry similar messages about the last act of creation by the gods. The inscription on Stela C from Quirigua, Guatemala (Figure 6), for example, reads:

> [T]hree stones are bundled, they plant
> a stone, Jaguar Paddler, Stingray Paddler
> it happened at First Five Sky, jaguar platform / throne stone
> he plants a stone [deity],
> it happened at Large Town? snake platform / throne stone
> and then it happened [he] bundled a stone Itzamnah
> water platform / throne stone, it happened at ?? Sky
> First Three-Stone place, 13 *baktuns* completed
> under his supervision, Six Sky *ajaw*[14]

Other Maya narratives speak of planting "three stones" (perhaps three stelae in three different places). These stones made up the Three-Stone Hearth of Creation—literally, the creation of the first hearth (contemporary Maya hearths often consist of three stones). According to some contemporary Maya groups, the hearth is symbolized by three stars at the base of our constellation of Orion (Kappa and Zeta, the easternmost star of Orion's belt, and bright Rigel). The fuzzy-looking Orion nebula, located at the center of the triangle, becomes the fire in the hearth (Figure 7a).

Stela 1 at Coba, Yucatan, reveals a flight of imagination on the part of one enthusiastic backward-directed scribe. A series of thirteens precedes the standard five-place Long Count. These numbers add up to a time base billions of years before our Big Bang creation, which modern cosmologists believe happened in a flash 13.7 billion years ago. The scribes' temporal meanderings remind me of contests my young playmates and I used to engage in when we would attempt to utter the largest number—jillions, zillions,

**6.** Stela C, Quirigua, Guatemala, is one of very few references to the creation day in 3114 BC in the Maya inscriptions. On that day the first three-stone hearth, the hearth of creation, was lit. No monument refers specifically to what will happen at the beginning of the next 5,125-year cycle. For the full text, see page 49. (Courtesy of Matthew G. Looper)

**7a.** The Three-Stone Hearth of Creation: Some modern Maya say the lower portion of our constellation Orion delineates the three stones of the creation hearth and the fuzzy-looking Orion nebula near the center represents the fire in the hearth. In many contemporary Maya communities the cooking fire is kindled over three stones (courtesy of Tyler E. Nordgren).

and gazillions! The rest of the text is worn away. Another stone carving, Monument 6 at Tortuguero, actually refers to the cycle-ending date in 2012, but the incomplete text that follows it does not specify what will happen when the deities mentioned in the text will descend.[15] Such Maya numbers are staggering, and I think it is easy to see why some Maya aficionados cannot resist the temptation to make the leap to universal meaning in these inscriptions, perhaps even interpreting such scanty information to foretell the second coming of Christ.

**7b.** The Three-Stone Hearth of Creation: The tortoise, an animal often connected with the act of creation probably because of the appearance of the swampy, watery world in which he dwells, bears the three hearthstones of creation on his back as he descends from heaven. (Madrid Codex, 71b; Akademische Druck-u. Verlag, Graz)

Scenes of world transformation also appear on the carved monuments, but they are just as difficult to interpret. As we will recall, much has been made by Y12 prophets of Stela 25 at the ruins of Izapa (Figure 2a). Recall that it shows a one-armed personage holding a staff and standing on the head of a caiman. Perched atop the staff is a giant bird connected to a ceramic pot out of which the staff emerges, and a long, thin serpent wraps around what looks more and more like a tree as your eye passes upward along the body of the caiman. Note that the rear end of the caiman seems to change into a series of leafy branches with blossoms at the top end. What are we to make of such a weird scene?

As we also learned earlier, the *Popol Vuh* tells us that an impostor sun lived in the world before the present creation. He was a bird deity with metal eyes and glittering gold teeth and was named Vucub Caquix, or Seven Macaw. From his lofty position high in a nance (breadfruit) tree, he boasted of his luster, saying "I am the sun; I am the maker of the earth."[16] Because the Maya believed modesty trumps self-importance, the gods sent hero twins to annihilate this faker. They shot him in the jaw with their blowguns; then, in disguise, they tricked him into thinking they could cure him (being clever is paramount in the Maya moral compass). Instead of providing Seven Macaw with sound dentures, the twins packed the vacant spaces with white cornmeal; they literally gave him a bad face job—a most deserving punishment for one who boasts about his good looks. Seven Macaw's teeth and his entire face fell apart and he perished.

As we know, Stela 25 along with a number of other artifacts (the Blowgunner Pot in Figure 2c, for example) likely depict this famous mythological scene, right down to showing one of the twins with a missing arm (in this story his limb was torn out of its socket during a ballgame with the feisty lords of Xibalba, the underworld). Like the Maya prophecies mentioned above, these vivid stories told in the many media are more about life's lessons and less about precise predictions in the cosmos concerning our collective futures. Clearly, this story basically relates to the evil of self-importance. Never brag about yourself lest you be done in by trickery.

John Major Jenkins bases a great deal of his argument about the galactic origin of the Maya Long Count on his interpretation of

Stela 25 and other iconography at Izapa. His theory also includes astronomical alignments at this early site. For example, he refers to the solstice alignment of Group F Ballcourt, the vicinity of which, he claims, displays creation imagery. Twenty years ago, I measured the axis of the Group F Ballcourt on which Jenkins places so much emphasis. Indeed, we found it to align approximately 1 degree off the December solstice sunset / June solstice sunrise direction.[17] However, the entire site exhibits that same orientation.[18] It seems a bit risky to pin all of one's conclusions regarding orientation on a single ballcourt at an early site—and a non-Maya, peripheral one at that.

We also determined that a number of Preclassic and Early Classic sites on the Pacific Coast were aligned solstitially. Figure 8 (bottom) shows the distribution in angle, measured from the cardinal directions along the horizon, of alignments of seventy-three cities spread throughout the Maya area. Note that there are two peaks in the chart. The smaller one lines up along an axis centered about 25 degrees off the cardinal directions. In the chart at the top of the figure, I have singled out the Preclassic site orientations. Although there are not many of them and original walls dating from that period that can be accurately measured are hard to come by, there is a noticeable clustering in the 20- to 30-degree zone. Measured relative to the east-west axis this is the direction that matches winter solstice sunrise and summer solstice sunset. The bigger peak, attributed largely to later period sites, may have more to do with a reformed version of the calendar based on solar zenith passages.[19] So there is evidence for a solstitially based calendar.

But which solstice? I think the June standstill of the sun makes more sense in calendars that originated in this area, because it marks a more important time of year in the agricultural year, the peak of the rainy season. The winter solstice date certainly cannot be ruled out, however, especially since it does happen to mark the Maya Long Count turnover.

What about the Milky Way analogy in the image on Stela 25 posited by Jenkins? Art historian Julia Guernsey Kappelman has

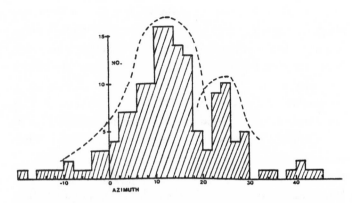

**8.** The distribution of alignments along the local horizon (measured from true north) for a large number of Maya sites from a variety of periods (*bottom*) shows two peaks. The smaller one (*right*) matches the distribution of Early and Preclassic Maya site alignments in southeastern Mesoamerica, where the calendar likely originated, shown in the top distribution. These data support the idea that the starting date of the earliest seasonal calendars in Mesoamerica may have fallen on one of the solstices. (A. Aveni and H. Hartung, "Water, Mountain, Sky: The Evolution of Site Orientations in Southeastern Mesoamerica," in *Precious Greenstone Precious Feather / In Chalchihuitl In Quetzalli*, ed. E. Quiñones Keber [Lancaster, CA: Labyrinthos, 2000]; drawing by author).

made a detailed study of the so-called Principal Bird Deity, who shows up in numerous early southern Mesoamerican sculptures, performing, flying, dancing, dialoging. She thinks he is part of a tradition related to the way the early rulers interacted with supernaturals.[20] In many instances the ruler impersonates the bird and is clad in a bird costume.

How many of us have had that recurrent dream in which we are flying? I wonder whether the ruler shared that deep-seated human desire to behave like Superman and take to the air, to fly up to the realm of the gods and commune with them. The twisted cord depicted on Stela 25 may be the cosmic umbilicus that connects all of humanity to the upper realm. Or it could be the ecliptic, the apparent annual path of the sun on the sky, shown in the banded imagery on page 74 of the Dresden Codex (Figure 5), which controls not only the movement of the sun but also the moon and planets, all of which regulate the calendar. As we recall, Seven Macaw has been identified in the sky as our Big Dipper. According to one interpretation he may be the agent who pulls the ecliptic northward so that the sun can reach its zenith, or overhead point, bringing with it the nurturing rain.[21] He may function as the alter ego of Itzamna, the God of the Sky. My main point here is that the imagery on Stela 25 focuses on the *ruler* and not some transcendent cosmic prediction.

As I hinted earlier, the available sources suggest that successive Maya creations are the result of the gods' attempts to bring about the ideal human race, the one that will tend the world, speak to the gods, and make the appropriate debt payment to them—the people who will keep the sun in motion and the world in balance. The Maya belief that there was more than one creation stands in stark contrast to the story of creation in the Old Testament, where it is portrayed strictly as a one-time affair accomplished by the spoken word of Yahweh. Also, the Maya creator gods are quite different from the God of Judeo-Christianity, and perhaps a bit more ordinary. They are anthropomorphic, that is, they take on the attributes of artisans, potters, woodworkers, and so forth; they are craftpersons who labor tirelessly to create the perfect race to people the earth. (Although in the second creation story in Genesis 2, God does mold man out of a handful of dirt.)

What of those earlier failed creations—mudmen, dwarfs, and monkeys? Vestiges of a few of them, such as the monkeys, are still around. They serve as a reminder of all the trouble the gods went through to bring us into the world. To clear the boards for their new and improved creations, the gods needed to purify the world, to cleanse it of their failed attempts. They chose to accomplish this through modes of destruction familiar to the people. That makes the story more interesting and relevant. But the gods needed to deploy these destructions on a grand scale, for our memories are short. The tale of destruction of the world by flood that appears in the Old Testament and on Dresden 74 reminds us that natural disasters are ubiquitous in the world.

My own studied opinion is that, like their parallels in the Bible, all Maya creation stories are designed to lay the foundation of the order of the cosmos, nature, and society and to point the way to achieve equilibrium among them. They are parables that refer to common knowledge shared by all Maya people, stories that tell of conflict between the good guys above and the bad guys who inhabit the world beneath us. Everybody knows these stories. They are fascinating, attention-getting tales told by parent to child around the fire. The Maya had their own core narratives, which involved the hero twins traveling along the Milky Way.

What did the Maya think about the Milky Way? There is no direct mention of the Milky Way in any of the Maya inscriptions. There is no "Milky Way table" in the codices, nor any identifiable Milky Way glyph. Spanish chroniclers tell us that the Aztecs called the Milky Way *Citalique*, or "She of the Starry Skirt," as well as a white road. I find no references to the Milky Way, however, in Landa's Maya chronicle; but the contemporary Maya also generally designate it as a road: the "road to the underworld," the "road of wind," the "road of rain," or, curiously, the "road of ice."[22] The origin of the idea that the Milky Way is a "world tree" and that its situation plays a major role in Maya creation originated in *Maya Cosmos*, by epigrapher Linda Schele and archaeologist David Freidel, one of the most popular books on Maya culture that emerged in the 1990s.[23] It relates a strictly contemporary account that following creation, First Father lit up the sky with the world tree, or the Milky Way.

At the heart of the Milky Way / world tree theory lies the notion that many images that deal with Maya cosmic symbolism are virtual maps of the sky, like the one shown in Figure 2b. The foundation hypothesis in *Maya Cosmos*, which clearly fueled Jenkins's galactic alignment theory (we will deal with that in Chapter 5), is that the Milky Way, portrayed as a tree, stood in the middle of the cosmos in perfect north-south alignment on the last day of the previous creation, as portrayed in the figure of the sky map. The key to deciphering the puzzle is the Blowgunner Pot mentioned above (Figure 2c). The scorpion positioned to the right of the base of the tree depicts the Maya zodiacal constellation of the Scorpion (amazingly, the same as our own Western Scorpius), which is located to the side of the Milky Way, although he faces the opposite direction in the sky. The bird at the top is the Big Dipper. *Maya Cosmos* is also the original source of the idea that carved monuments, such as Izapa's Stela 25, depict that very alignment.

There are huge problems with many aspects of this theory. First, the galactic alignment in the sky is far from unique. Art historian Susan Milbrath was first to point out that most of Freidel and Schele's arguments about how the Milky Way lines up may pertain to seasonal events, rather than to events that occur over vast epochs.[24] Thus, the celestial reenactment of the creation of the world could refer to a ceremony carried out once a year rather than a one-time creation event. As a parallel, think of the way Christians celebrate the life of Christ by reenacting the birth of Christ around the time of the winter solstice, along with seminal episodes in their savior's life between Christmas and Easter, the resurrection. In fact, the Milky Way actually aligns north-south every night of the year at one time or another, or at specific times of the night on various dates in the seasonal year.

A second and deeper problem with the Milky Way / world tree theory has to do with broader issues about mapping in general and especially our ethnocentric insistence on using maps as forms of expression. I want to elaborate on this issue here because I believe it is distinctly related to ways we often misinterpret the Maya. When Schele and Freidel inform us that what works in the sky matches the creation condition, they seem to be saying that "what works" functions like a map in our culture. Trouble is, nowhere in the Mesoamerican record is there any indication that a representational device that consists of a flat surface displaying in scaled proportion the relative position of selected points—that is, a map—ever existed. I have argued elsewhere that maps, as we know them, are *exclusively* a product of Western culture, descended from the Greeks.[25]

The closest Maya document I know of that resembles a map appears on pages 75–76 of the Madrid Codex, a pre-Columbian codex from fifteenth-century Yucatan (Figure 9). Let me introduce this fascinating cosmogram with a riddle: What do the Maya and Albert Einstein have in common? Answer: Both sought harmony by attempting to combine space with time. Madrid 75–76 shows the four sides of space (not the four cardinal points as we reckon direction) depicted as areas encompassed by the flaps of a Maltese cross. Each flap of the cross contains the things that inhabit that direction: gods, colors, parts of the body, flowers and trees, and even days. Time envelops space. You can count it off via the 260 dots that make up the periphery of the world.

Only after Hispanic contact do we begin to see what look like our Western maps in the literature, although there is still something "native" about early Yucatecan colonial maps: neighboring towns are plastered equidistantly along a now circular horizon centered on whatever town serves as home base. All of these native maps are *loco*-centric, not *helio*- (sun) or *galacto*-centric. Their message seems to be that what really matters in the universe can be discerned in a reference frame centered on where we live, not on some far-away place in the universe, as our modern cosmological maps portray the situation.

Maps from the Mexican highlands, such as those from Cuauhtinchan, near Puebla, Mexico, depict space as a journey made up of meandering footprints marked by time notations.[26] Mountains border the maps—again all positioned equidistant from home base. If you place one of our modern geographic maps, with its lines of latitude and longitude, next to it, you will scarcely find anything accurately placed. In the Mesoamerican world view the distance to one's destination is measured by the time it takes to get there. Ask

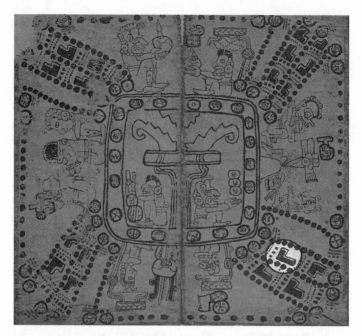

**9.** The Maya view of space-time is depicted in a cosmogram on pages 75–76 of the Madrid Codex. It pictures four world directions, each as a flap of a Maltese cross. Each region contains its own set of gods. The 260-day *tzolkin*, one dot standing for each day, envelops the world. Footprints (one of them is circled) along the road one travels in time lead to the fire at the center. (Akademische Druck-u. Verlag, Graz, and Museo de America, Madrid)

any itinerant peasant walking along the side of a road, bearing a load of wood on his tumpline: "Which way to Xlapak?" and he will tell you "Adelante" (ahead), "poco," or "mucho" or indicate to this or that side with a gentle motion of the hand. Face it: then and now theirs is a less graphic way than ours of representing space. The Maya are different from us. Rejoice in the insistence of diverse indigenous customs!

The same is true of star maps. The pre-Columbian Maya had none. Some colonial star maps drawn by Spanish chroniclers, who interviewed native calendar keepers in highland Mexico, have survived. But they do not resemble Western star maps in the slightest; for example, Orion's belt and sword, which make up the Aztec constellation of the Fire Drill, do not exhibit the correct number of stars. And the stars that make up the Pleiades, known to them as the Rattlesnake's Tail, are rendered in impossible positions, encircled by a chain of stars that does not even exist.

Now, I can well imagine why someone living in our modern era might think the Maya made maps. After all, we have a habit of using a two-dimensional window on our monitors to conveniently download a map of the sky as it appeared at any time and in any place from a wide selection of astronomical software packages. It is easy to change our sky into theirs. Half a century ago astronomer Gerald Hawkins achieved instant fame when in his best-selling *Stonehenge Decoded*, he used the then-novel computer to deduce that England's celebrated megalithic monument was in fact a computer.[27] One of his critics, historian Jacquetta Hawkes, was moved to utter the oft-quoted quip, "Every age gets the Stonehenge it deserves—or desires."[28] The idea of sky-as-template has emerged as a full-bloom contemporary paradigm in cultural astronomy.

I will give two brief examples here (and discuss them in detail later). In *The Origins of the Mithraic Mysteries*, philosopher David Ulansey reinterprets the Mithraic myth of the slaying of the bull as a map of the zodiac on the day of creation.[29] And Robert Bauval, in his *The Orion Mystery*, argues that if you position the Milky Way over the sands of ancient Memphis parallel to the left bank of the Nile, you will find a constellation template that fits the location of the pyramids.[30] His imaginative scheme equates the size of each pyramid with the brightness of the corresponding star in the template that represents it. I wonder just how the Egyptians might have managed that. And I wonder too whether, to paraphrase Hawkes, the Maya map-questers might not be giving us the twenty-first-century Maya we deserve.

I highlight all of these examples to remind us that it is all too easy to project our own habits, attitudes, and customs onto the culture we are studying. Sure, the Maya and the Aztecs cared a lot about precise celestial periodicities and astronomical alignments, but there is no evidence that portraying astronomical phenomena in chart and map form, as is our custom, held any interest for them. The same holds for the roundness of the earth or the center of the solar system. Why force our ideas on them when we can better spend our time listening to their ideas? Maps just were not part of their way of expressing things.

My third and final basic problem with the Milky Way / world tree theory of creation has to do with sources of evidence. We need to be especially aware that all Maya Milky Way identifications come exclusively from contemporary Maya people. There is no evidence that what they tell us bears any connection to what Classic Maya astronomers saw in the sky 2,000 years ago when they concocted their Long Count calendar. Moreover, there is no account that I know of in any of the written literature from colonial times in which the Milky Way is described as a tree.

Well-informed scholars may not agree on all aspects and details about what basic Maya symbolism represents, and given such scant evidence how can we possibly comprehend what the Maya really thought about 2012? Although there is not complete agreement about what aspects of the sky pertain precisely to the story of creation, this much is clear about the Maya philosophy of cyclic time: it is past-oriented. Maya creation is far from a chain of non-repeatable progressive events pointing toward a Biblical-type Armageddon, the way so many Y12 prophets insist on viewing 2012. Our time will come and we will have a chance to redeem ourselves before we return to earth anew—so say the contemporary Maya. And I guess that is what appeals to so many of us about Maya cosmology, especially given the atmosphere of gloom-and-doom fatalism that seems to come from so many interpretations of scientific accounts of the Big Bang and evolution by natural selection.

Modern science has taught us that there is little we can do to influence the course of nature on a grand scale. Our spirit of scientific inquiry has constructed a universe that offers, for many of us, too little room to participate. This was not so in Maya cosmology and I think that is one reason why we envy them. In the Maya way of thought the cosmic connection between humanity and the sacred gets reestablished continually through reenacting the creation story at the termination of all Maya time cycles, great and small. This the ancient Maya accomplished in their rituals of debt payment to the gods. There is no differentiation between myth and history in the Maya mentality. Cosmic and human time merge into a kind of "mythistory" every time a cycle overturns.

To sum up, our knowledge of ancient Maya culture suggests that long period cycles carved in stone served the purpose of demonstrating that the rulers were the very embodiments of their ancestors who lived in previous eras and now reside in the afterworld of the sky. Carved Maya stelae were a form of political and religious esoteric propaganda, a combination of what comes across to us as both myth and real history devised expressly to advertise Maya political doctrine. Like the Egyptian hieroglyphs, the signs carved on the Maya monuments record seminal events punctuated by specific time intervals in the life of the ruler—ancestral affiliations, accession to office, captures, marriage, childbirth, and so on. This is the interpretation that fits best with what we can now read on Maya monuments like Copan's Stela B. The ancient Maya were far from the futurist egalitarian society many modern day prophets of 2012 imagine they see in the world's ancient civilizations they choose to romance.

The inscriptions are not about prophecy in a literal sense; they were written to reinforce political power and to maintain stability. If you want to connect with cosmic forces to justify your actions, as the Maya rulers surely did, then you pay close attention to the movement of celestial bodies—their times of appearance and disappearance, their coming together, their occasional awkward turning backward. You monitor the traffic on the road of the sun, moon, and planets—the zodiac. You train specialists to develop the complex mathematics necessary to predict when events in the immediate future will take place in the heavens. There is sound religious and political capital in knowing what is going to happen next through skywatching.

Next we turn our attention to what was perhaps the most artful and exquisite Maya achievement—the calendar. As outsiders we are dazzled by the magnificence and enduring accuracy of their timekeeping devices and the dedication of those who devised them. But again, we will need to keep in mind the social context in which Maya calendars (for there were many of them) and their attendant cosmology developed. Like our Big Bang cosmologists, the ancient Maya daykeepers, although charged by ruling kings and queens, discovered cycles so grand that they defied comprehension. What they achieved subverted divine cosmic forces to their own ends—sustaining power. This was the message and there is little doubt about the audience to whom it was directed—the Maya themselves.

# THE CALENDAR:
# JEWEL OF THE MAYA CROWN

Western civilization is not alone in seeking its origins in deep time. We bundle our years into decades, our decades into centuries, and our centuries into millennia. Our ages—the Age of Reason, the Age of Enlightenment, the Middle Ages—are packaged into eras, such as the Christian and pre-Christian eras. For the believer, the Christian era will end with the second coming of Christ, for in the Christian historical view all things were made by God expressly for the ends they fulfill. The new era that will follow will consti- tute a timeless eternal existence to be experienced only by the true believer. Philosophers call such a temporal concept a teleological timeline, because it is dictated by things that happen at the end, which are responsible for propelling time's arrow forward.

Before Christianity introduced this linear concept, "big time" in the West was based in the pagan tradition of the Classical world. Time was made up of rhythmic, repetitive events centered on the return or reenactment of earlier events often reckoned by celestial cycles, such as planetary conjunctions. (Recall our definition of the two kinds of time in the Preface—historical-linear and mythic-cyclic.) Crossings of Jupiter and Saturn were popular choices in the ancient Chinese calendar, whereas the Chaldeans of the Middle East favored the assemblage of all the visible planets in the constellation of Cancer. The Hindu calendar, on the other hand, was a purely mathematical contrivance based on 1,000-year multiple cycles of years, called *yugas*. The grandest cycle of time measured in *yuga* lengths was thought to be a "day" in the life of Brahma. The bigger the tree, the deeper the roots. One way or another, all complex civilizations ultimately establish their origins in the very distant past.

The Maya were no different when it came to the subject of time. They wove the history of their dynasties into the fabric of deep time in order to legitimize their right to rule. Our contemporary political leaders do no less when they conjure up famous figures from the past as role models: if it was good enough for Lincoln, or Washington, or Reagan, or Roosevelt, then it's good enough for me! The Maya ruler also took advantage of time's natural indicators in the sky as vehicles for validating authority. I do not mean to suggest here that the king or the commoner tilling the fields did not hold to any fundamental set of deeply revered beliefs underlying a well-thought-out Maya philosophy of time; in other words, I do not believe the power structure was simply manipulating time to hoodwink the people. That is too simple.

During the Classic period the Maya developed a passionate interest in time and number. I think this is one of our biggest reasons for admiring them—they seem so much like us. By the middle of that period their interest flowered into a fascination that bordered on obsession. It is as if scribes and calendar keepers, all members of the elite class, perhaps led by one or two unknown geniuses, the likes of Newton and Einstein, had created a veritable Maya Institute of Advanced Studies. By examining some of the inscriptions the Maya produced during this exciting intellectual period we can begin to acquire a feeling for this mathematical passion and the skill that accompanied it.

Recall from our translation of Copan's Stela B (Figure 4) that the first bits of information in a Maya inscription consist of numbers that refer to time. What distinguishes the Maya love affair with numbers is their preoccupation with what I have called the commensuration principle—the habit of organizing time cycles, large and small, to interlock and fit together in ratios of small whole numbers, such as eight to five, the seasonal year and the Venus cycle. (We will discuss some of these specific ratios below.) Where did these ideas about time management come from and how is it that timekeeping was catapulted to such a lofty level in Maya culture?

Although what survives is carved in stone, the Maya probably engraved their earliest chronological records in wood. The idea of encapsulating historical events in a closed chronological network of time loops likely came from the Olmec culture, their antecedent neighbors to the west. They also borrowed from the Zapotecs, who lived farther to the west in the highlands of central Mexico around the region of Oaxaca. Pre-Maya monuments from that region, dated possibly as far back as 600 BC, exhibit some of the earliest references to timekeeping in the New World. These rudimentary hieroglyphs are a far cry from the ornate forms we find at places like Copan, Tikal, and Palenque during the Maya's Classic period. Still, the Maya shape of time clearly begins to coalesce in them.

The way some of these early numbers are written offers clues to how the Maya's pre-numerate ancestors once used the parts of their bodies to count the days. Round circles or dots stand for ones and bars represent fives, as shown in Figure 10a. (You can also find them above the pictorials in Figure 3.) I think the dot symbols are abstractions that probably represent the tips of the fingers, and the bars are the extended hand with fingers closed, gestures used by pre-Maya antecedents for tallying. Indeed, we still speak colloquially of handfuls of things. If you begin with the little finger of the left hand (call it day number one), then count across both hands through ten, and then across the toes to twenty, you literally will compile a "person-full of days"—about three of our weeks. This is an easily cognized short duration. The completed body is often represented by a closed fist or stylized conch shell that resembles it. This symbol of completion stands for zero. (Examples are shown in Figure 10b.) The Mesoamerican counting system operates the same way as our decimal (base-10) system, except that it takes all the fingers and toes, rather than just the fingers, to fill a position in a number sequence. Twenty is the basic unit in the Mesoamerican counting system; no rational combination of dots and bars in a single position has ever been found that exceeds that amount.

The Maya revered the base-20 numbers that made up their vigesimal system to such a degree that they fancied each of them a god. In many Maya inscriptions a defining head, or in some instances the full-body figure, of the god portrays the number instead of the simple dots and bars. Often number deities on stelae are depicted bearing the burden of time, which they carry in their backpacks along the road of time (Figure 10c, right). They deposit their load of time at our feet as we face the monument. Thus, time is just like one of the commodities borne by merchant travelers.

On all the stelae that have been deciphered, the fundamental unit of time is the day. Contemporary Maya still call it *k'in*, a term that also means "sun" and "time." The Maya conceived of the day as a direct manifestation of the annual cycle of the sun. In other words, time *is* the sun's cycle itself. The hieroglyphic signs for *k'in* are among those most frequently displayed in Maya writing (Figure 10d shows two examples). The tips of the floral symbols at the center of each cartouche or frame, which stand for procreation, may signify the extreme positions of the sun at the horizon, where it rises and sets at the winter and summer solstices. The earliest records of signs and symbols that resemble day names in Mesoamerica emanate from the middle of the first millennium BC, specifically from the area around Oaxaca, the Gulf Coast, and the highland Mexican sites of Cuicuilco and Chalcatzingo.

The Maya built their cycles of days into "months," or *uinals*, and they gave each day in that twenty-day sequence a name—usually that of an animal or force of nature, such as jaguar, monkey, wind, and night. (For some examples, see Figure 10e.) By 300 BC, some would argue possibly as early as 600 BC, the Maya had acquired a system of counting days in a bigger cycle—one that measured 260 days. The complete cycle, called the *tzolkin*, or "count of days" and the sacred round, was probably invented by pairing, or "commensurating," two smaller cycles: number coefficients one through thirteen (the number of layers in Maya heaven) and the cycle of the twenty day names. The day 4 Ahau is the *tzolkin* date in the inscription on Copan's Stela B. We do the same thing when we set the thirty (or thirty-one) numbered days of the month alongside the cycle of the seven day names of the week.

There is nothing quite like the 260-day cycle anywhere else in the world. The *tzolkin* is the centerpiece of the Maya calendar system and the hallmark of the principle of commensuration in the Maya calendar. It is the single most important chunk of time the Maya ever kept—and still do keep in areas remote from modern influence. But why 260? A number of theories have been put forth to solve this mystery.

Was the 260-day time count born simply out of multiplying the numbers thirteen and twenty? Some experts think so. Although we know that thirteen may be basic because it represents the number of layers in the Maya upperworld, there is some disagreement about whether this was a valid concept in pre-contact times. Or could 260 have emerged as a seminal number because it connotes something natural in human experience? Biorhythms offer a possible answer. The average duration between human conception and birth is 266 days. Today Maya women still associate the *tzolkin* with the human gestation period. They time their term by the moon, counting nine months of the phases—265.77 days by modern calculations. In some parts of Yucatan they still say that the moon draws "nine bloods" away from the pregnant mother to give to her newborn. Furthermore, the birthing cycle is a fair approximation to the length of the basic agricultural cycle in most areas of the Maya world. So 260 neatly ties two fertility cycles together, those of woman and earth.

**10.** Here are some examples of Maya time notation:

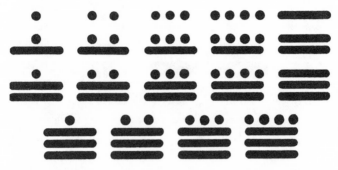

**a.** Dot and bar numbers one through nineteen. Likely descended from pre-numerate hand gestures, bars stand for fives, and dots for ones.

**b.** The zero symbol is often represented by a closed fist, signifying completion, or by a shell, as shown in these examples.

**c.** Each number is also represented by a deity, the face or, occasionally, the full body of which can be used instead of the dots and bars. *Above:* head variants one, two, and three; *below:* the full-body gods fifteen and five carry their own bundles of time. Note the facial resemblance between teen and ordinal.

**d.** The *k'in* glyph, two examples of which are shown here, stands for sun, day, and time.

Etz'nab

Zotz

Cauac

Tzec

Ahau

Xul

**e.** Day and month signs. *Left*: three of the twenty day names of the 260-day *tzolkin*; *right*: glyphs representing three of the nineteen months of the 365-day *haab*. (From A. Aveni, *Skywatchers: A Revised and Updated Version of Skywatchers of Ancient Mexico* [Austin: University of Texas Press, 2001], figs. 51a, 51b, 53, 56, 58; A. Aveni, *Empires of Time: Clocks, Calendars, and Cultures* [Boulder: University Press of Colorado, 2002], fig. 6.3; drawings by P. Dunham)

I have long suspected that there may be astronomical reasons behind the origin of 260. First, the average interval of the planet Venus's appearance as morning or evening star is 263 days.[1] Second, the average duration between successive halves of the eclipse season, 173.5 days, commensurates with the *tzolkin* in the perfect ratio of three to two. If this seems contrived, there is evidence in the inscriptions that the Maya used the *tzolkin* to predict when Venus would appear and when eclipses would occur. For example, certain named days in the 260-day count were tagged as unlucky because they marked a period of inauspicious events, such as eclipses. Considering that their society was so carefully attuned to anticipating celestial events that signaled temporal transitions in their lives, we can only imagine the drama attending an unscheduled plunge into darkness at noon. In the Maya Dresden Codex a table of eclipse warnings names these inauspicious days and associates omens with them that deal with human pregnancy and the maize cycle. The three-to-two commensuration between the eclipse cycle and the *tzolkin* guarantees that certain days particularly vulnerable to the occurrence of eclipses will fall in clusters at intervals one third of a cycle (about 120 days) apart in the *tzolkin*.

A third celestial rhythm with a 260-day beat that has meaning only in tropical latitudes indirectly has a bearing on popular ideas about 2012. It is connected with the number of days the noonday sun spends north or south of the overhead position or zenith. These intervals vary depending on latitude, but in latitude 14.5 degrees north, close to the locations of the great Maya city of Copan and the peripheral site of Izapa, the annual cycle divides neatly into 105- and 260-day periods. (Recall Izapa's largely unproven connection with the origin of the Long Count.) Based on the present-day archaeological record, however, the "ideal latitude" lies a bit on the periphery rather than at the center of the area where archaeologists have unearthed the earliest calendrical inscriptions. Also, unlike the northing and southing of the sun, the *tzolkin* is not fixed in the seasonal year. It just rolls along, beginning a new cycle where the previous one ended, regardless of where that point fell in the year measured by the seasons. Finally, timing a zenith passage precisely is no mean task, for the sun is an extended object like a star, rather than a point. A two-day error is easy to make. What it all boils down to is that even the most careful skywatchers could confine the event only to a band about 100 miles wide centered on latitude 14.5 degrees—which covers a lot of Maya territory.

So where did 260 come from? My best guess is that the sacred count of days acquired its importance when some enlightened Maya daykeeper realized that the number 260 brought together many things. We can compare this magic number to our gravitational constant or the speed of light—numbers that repeatedly assert their presence in so many mathematical calculations in both classical and modern physics. In my opinion the discovery of this grand commensuration—the harmonic focal point of so many of nature's constructs and phenomena, such as human anatomy, birthing, the moon, Venus, and eclipses—likely did not arise in the number-oriented heads of Maya daykeepers all in a flash. But with the Maya focused so intently on the idea that nature and number are joined together perfectly, the discovery of the multiple significances of 260 was bound to be raised to prominence in Maya time consciousness.

Regardless of its origin, the *tzolkin* was, above all, a cycle intended for divining, for communicating with the gods. In the almanac from the Madrid Codex pictured in Figure 3, for example, the sky god and death god receive offerings (they appear to be headdresses or perhaps breastplates) from supplicants. You even took your name and your fortune from the day name in the 260-day count on which you were born; for example, 1 Imix ("One Jaguar"), or 9 Oc ("Nine Dog"). And modern diviners in highland Guatemala still pass through a 260-day initiation period before they are allowed to practice. Given the depth of penetration of the *tzolkin* into Maya notions about time, I should think that anyone who seriously proposes that the Maya came from Asia or Africa would find the number 260 as basic to those supposed source cultures. That I have seen no such argument has only bolstered my anti-diffusion orientation on this issue.

From descriptions of the *ah k'in*, or daykeeper (see page 39) in the chronicles, we are already aware of the high social status and extraordinary skills the court astronomer-mathematician-priest possessed. He or she was a master of scientific computation and a learned practitioner in the art of calendrical divination. Descendants of such experts still exist among the Maya today. A few of these calendrical diviners claim to have retained a knowledge of the Maya cycles of time and they appear to be generally aware of the way these rounds of time link together.

Modern anthropologists have observed that day counting and tallying is but a small part of the repertoire of these specialists, who are still held in high regard by the indigenous community. A diviner's task is not simply a matter of consulting a list of days and naming their properties. The whole process of making calendrical prognostications operates more like a dialog between priest and client, and much of the outcome depends on their social rapport. Anthropologist Barbara Tedlock has described one such dialogue. Here the diviner announces that he is taking hold of the divining bag and borrowing the health of the particular day (of the 260-day cycle) on which the divination is taking place:

> "I am now borrowing the yellow sheet-lightning, white
> sheet-lightning, the movement over the large lake, little lake,
> at the rising of the sun (east), at the setting of the sun (west),
> the four corners of the sky (south), the four corners of the
> earth (north)." At this point, sensing that the "blood" and
> the days are ready to respond, the diviner, after saying "one
> is now giving clean light," then proceeds to frame the divina-
> tion in a formal way. For example, the first formal question in
> the case of illness would be, "Does the illness have a master,
> an owner?"[2]

The calendrical divining process in ancient times may not have been so different, except that the *ah k'in* probably carted along a codex, or book of computations and divining, for detailed consultation in various towns.

As I mentioned earlier, our misfortune is that practically all the books that might offer us insight into just how the ancient divining process worked were destroyed in huge bonfires set by the Spanish priests, who feared that Maya books promoted idol worship. "We found a large number of books written in these characters [hieroglyphs] and, as they contained nothing in which there were not to be seen superstition and lies of the devil, we burned them all, which they regretted to an amazing degree,"[3] wrote Friar Landa. He then provided an inventory that included numbers of idols and pots smashed, monuments toppled, and so forth. "Know thine enemy" is good advice for any conqueror. Conversion is a connection point: what you convert *from* is as important as what you convert *to*.

In addition to the *tzolkin* cycle of 260 days, the Maya also kept a seasonal year cycle known as the *haab*. It consisted of eighteen named months, each of twenty days. In our earlier reading of Copan's Stela B, 13 Yax (the thirteenth day of the month Yax) is the *haab* date. Calendar keepers added a nineteenth month made up of five so-called unlucky days, to round out the 365-day year count. Curiously, the Egyptians did something similar: they tallied twelve months of thirty days and tacked on five days of misfortune at the end. I think the unluckiness tied to these days in both calendars may stem from the notion that they lie outside of time's order. There is harmony in "commensurateness"—it satisfies.

Slippage between the 365-day year count and the actual year of 365.2422 days, measured by the annual course of the sun in the sky, did not seem to matter to the Maya. They did not on certain occasions add days to the year count, the way we do with leap years to keep our holidays from sliding backward through the seasons. What if Christmas retreated into autumn or the Fourth of July backed up into the cold of winter? For reasons we will probably never know, the Maya seemed to place more emphasis on following an unbroken chain of time, as they did with the *tzolkin*. They were different from us.

If the whole idea behind timekeeping in complex societies is to extend the past and anticipate the future, then the more organized and expansive a culture becomes, the more motivated are its leaders to devise bigger and bigger cycles. The next largest cycle in the Maya calendar is formed out of a commensuration between the 260-day *tzolkin* and the 365-day *haab* cycles. The Maya "calendar round," as scholars have chosen to call it, is fifty-two years, or 18,980 days, long. This is the lowest common multiple of days in both the *tzolkin* and the *haab* (52 × 365 = 73 × 260 = 18,980 days). Thus, the calendar round is the interval over which name and number combinations in both the *tzolkin* and *haab* cycles repeat themselves. For example, 4 Ahau 8 Cumku, the day of creation in the Long Count, will recur every 18,980 days.

There is evidence that the completion of a fifty-two-year cycle, which amounted to a fairly hefty lifetime then, was celebrated all over ancient Mesoamerica. For example, in the New Fire ceremony, sort of a mega New Year's celebration, the new beginning is timed by events in the sky and is heralded in an oft-quoted chronicle of the Aztecs. The site is a temple on a low platform located on a mountaintop, called the "Hill of the Star," just outside the capital city of Tenochtítlan (Mexico City). There calendar keepers gathered at midnight to mark the event:

> And when they saw that [the Pleiades] had now passed the zenith, they knew that the movements of the heavens had not ceased and that the end of the world was not then, but that they would have another 52 years.[4]

The chronicler continues,

> Behold what was done when the years were bound—when was reached the time when they were to draw the new fire, when now its count was accomplished. First they put out fires everywhere in the country round. And the statue, hewn in either wood or stone, kept in each man's home and regarded as gods, were all cast into the water. Also (were) these (cast away)—the pestles, and the (three) hearth stone (upon which the cooking pots rested); and everywhere there was much sweeping—there

**11.** An image from page 34 of the central Mexican Codex Borbonicus depicts the completion of a fifty-two-year cycle being celebrated in a New Fire ceremony. Note the resemblance between the completion glyphs that appear in this scene and the cosmogram in the Madrid Codex shown in Figure 9. (Akademische Druck-u. Verlag, Graz)

> was sweeping everywhere. Rubbish was thrown out; none lay
> in the houses.[5]

This was no simple housecleaning, but rather a ritual act of purification that marked the beginning of a new cycle of time for the Aztecs. A New Fire event is pictured most vividly on a page of the Borbonicus Codex, dating to the fifteenth century (Figure 11). Bearers of reed bundles come from the four directions to feed the ceremonial fire at what will become a symbolic new three-stone hearth. Completion glyphs mark their eyes and the temple doorway, where they make their offering. Note that the shape of this glyph resembles the space-time cosmogram in Figure 10d.

By the second century AD, Maya polities had fully mastered culti-
vation of the land, built great cities, erected exquisite monumental
architecture, and expanded the state. A few hundred years earlier,
the rulers had created the Long Count, a brilliant invention fashioned out of a huge buildup of base-20 cycles (described in more
detail below). This fundamental innovation in their calendar was
used by Maya royalty to fabricate a dynastic narrative that covered
vast stretches of elapsed time and held the potential to extend the
taproot of Maya culture all the way back to the creation of the gods
themselves.

Monumental inscriptions always begin with an event, then an
interval; then another event follows, another interval, another event,
and so on. Most of the events in the inscriptions denote the births
and deaths of the ruler's ancestors and their connection with seminal points in his/her life. Epigraphers call these intervals "distance
numbers" to convey the Maya way of thinking of time as distance—a
road traveled. Things happen at the rest periods, or breaks, separated
by these distance numbers. That is when the lords of number who
walk the road of time lay down their burdens, as we remember from
Figure 10c. All events are pegged to a Long Count date that appears
at the beginning of the inscription. It places the opening event being
commemorated in a time count dating from creation. Inventions like
the Long Count and distance number allowed the ruler to proclaim
the extraordinary longevity of his bloodline in concrete terms.
Suddenly his ancestry acquired great depth. The commoners who
stood in front of his stelae and read the text (or, more likely, had it
read to them by an official because they were illiterate) acquired a
real feeling of the ruler's power and permanence.

Because it plays such an important role in 2012 issues, let us
look at precisely how the Maya Long Count operates. You can compare it to the odometer on your car, except that instead of tallying
miles, the Long Count clicks off one day at a time in endless succession. This analogy can be a bit misleading, because there actually
is no evidence that the Maya ever used gears or machinery to keep
time. Regrettably, there is another essential difference between your

automobile and the Maya universe of time. When their "odometer" turns over, thus signaling the resting point of the longest Maya time cycle of all, the cycle begins anew. By contrast, your automobile just gets older as it heads ever closer to the junkyard.

Not surprisingly, Long Count inscriptions operate in a base-20 system; in other words, each place in the series of numbers that makes up a Long Count date contains twenty times the quantity of the previous. Thus, twenty *k'ins* equal one *uinal*. The exception is the third place upward in the hierarchy, the *tun*, which means "stone"; it holds 18 times 20, or 360 days, instead of the logical 20 times 20, or 400 days. This difference is probably because 360 days is a closer approximation to a year than 400; therefore, a quick glance at a *tun* count immediately gives you a rough idea of how many years it represents. Interestingly enough, the Maya used units of 400, 8,000, and so forth in their trade count, that is, when they counted things, such as cacao beans. So, as in our system of numeration, wherein the cycles of seconds and minutes overturn at 60 instead of 100, time merits a special counting system.

Some clever Maya mathematician extended the *tun* cycle by multiplying each successive order by twenty. In a sense, that genius flattened out a portion of time's circle to lend it a more linear appearance. To give an analogy, imagine looking out over a distant horizon. Experience tells you the land before you is flat, but you have been taught that if you travel far enough, your trajectory will curve back on itself. Likewise, it is only when we contemplate the Long Count over extraordinary intervals, like bundles of years instead of days, that time becomes cyclic again. Thus, twenty *tuns* made up a *katun*, or 7,200 days; and 20 *katuns*, a *baktun*, or 144,000 days; thirteen *baktuns* make one creation period, which amounts to 5,128.77 *haab* or 5125.37 seasonal years. Normally the *baktun* is the highest number appearing in the chain, but some computations in the codices creep upward a few more cycles. (Recall that Stela 1 at the ruins of Coba, Yucatan, prefixes nineteen more multiples of thirteen, thus catapulting a king's putative ancestry all the way back to well before our Big Bang creation. (See page 49.)

To illustrate how the Long Count works in practice we will read the time text on Copan's Stela B (Figure 4). It begins with 9.15.0.0.0, or $(0 \times 1) + (0 \times 20) + (0 \times 360) + (15 \times 7,200) + (9 \times 144,000)$ days. This adds up to 1,404,000 days since creation, which translates to August 20 (or 22), AD 731 using the accepted correlation between Maya and Christian calendars.[6] We noted earlier that Stela B is a *katun*-ending monument, that is, its last three places read zero. The inscription that follows the Long Count advertises the fact: "[T]here were completed 15 katuns." This is something like our erecting a monument every time a decade or a century elapses (a *katun* is approximately a score of years).

We know roughly when and why the Long Count was set up. But we do not know how; that is, we do not know how the start date for the Long Count was chosen. Did they back-calculate to get to it? If so, how? Here I think we can take a lesson from the way other civilizations have accomplished this task. Our own "long count" is a good example because we have a lot of evidence about how it was established. Our calendar, of course, is reckoned from the putative date of the birth of Christ. The tally of serial years was not worked out until the sixth century AD, by the monk Dionysius Exiguus ("Dennis the Short," literally), in the Eastern (Christian) Holy Roman Empire. It is worth noting that precise timekeeping in the West was first the business of religion rather than science. People needed to know when their prayers would be most effective. Prior to that, chronologies were expressed in Biblical "begats" and lists of kings, systems that were both confusing and not computationally friendly. (Nevertheless, that is how Irish theologian and scholar Bishop James Ussher [discussed in greater depth in Chapter 6] arrived at the still-popular 4004 BC creation date of the world, that is, by counting the generations in Genesis back to Adam and Eve.) Dionysius's goal was to mark time's zero point by Christ's birth as reckoned since the foundation of Rome, then thought to be December 25, 253 AUC ("since the foundation"), but curiously he started the count on January 1, 254. That he was also wrong about Jesus's birthday makes Dennis the Short's calendar even more inaccurate. The sequence of years BC, or before the birth of Christ, was not initiated until more than a thousand years later, in AD 1627 by the astronomer Denis Pétau (aka Dionysius Petavius). Even with its badly flawed zero point, this abstract, rational arithmetic scheme still serves as the framework we use to order all historical events in the long term. Think of it as history's time machine.

Imagine Little Dennis, like Bishop Ussher, laboring to set time zero. Imagine, too, some astute Maya *ah k'in* sitting in his temple and scribbling calculations on bark paper to figure out how his ruler's lineage might fit into some already-canonized information about the gods and creation. Since most carved stelae date from *baktun* 9, with just a few dating to *baktun* 8 and even fewer to *baktun* 7, it is likely that all this calculating to set up the Long Count happened some time before the Classic period (AD 200–900), just as our slightly erroneous BC/AD system of counting was instituted well after Christ, whose nativity marks its zero point.

Thanks to some valuable sixteenth-century documents found in Yucatan that clearly identify dates in the calendar round with their equivalents in the Christian calendar, we can translate the Long Count into time as we measure it.[7] When we do, we discover that day zero of the last Maya creation fell on August 11 (or 13), 3114 BC (or –3113 in a system that includes the year zero). When we march forward 13 *baktuns*, or 5,125.3661 years, we arrive at day zero of the next creation: December 21 (or 23), 2012. Now it may be pure coincidence, but the August 11 day comes close to the day of solar zenith passage in southerly Maya latitudes, in the general region where the Long Count calendar got started; and December 22 (give or take a day) is the winter solstice (or solar "standstill"), which marks the day the sun arrives at its maximum southerly position in the sky. It is conceivable, then, that the two creation events are keyed to important positions of the sun cycle. (Let's do the math: the number of days in 13 *baktuns* is 1,872,000 days. If we divide that number by the number of days in a seasonal year (365.2422 days), we get the length of one creation period: 5,125.3661 years. The remainder, 0.3661, equals 133.7 days. The number of days between August 11 and December 20 is 133 days, so it is pretty close (more on these two seminal dates in the next chapter).

I noted earlier that, as observable phenomena, neither solar zenith passage nor solstice are easy to establish in fixed time. You can use a gnomon, or vertical shadow-casting stick, to target zenith passage within a day or two. (Recall that because the sun is an extended object, its shadow is diffuse.) Also, the sun's motion slows dramatically as it approaches solstice (the sun moves less than 1/10 of its own diameter over a six-day period spanning the solstice). This means that assigning a time to the precise place on the horizon where it rises or sets can be incorrect by several days. Of course, the Maya could have counted the time it took for the setting sun to move from a discernible pre-solstice setting point to the solstice and back again to that same point; they could have divided that time interval by two and then added it to the first date to get the solstice. The problem is that we do not know that they did. I am not just picking nits here: if you are going to base your theory of the calendar on astronomical time periods, you need to have an observational base. You need to *see* the phenomenon to get to the numbers behind it. (I will have more to say on this point and other astronomical considerations in the next chapter.)

So much for the seasonal dates. But why 3114 BC? Anthropologist Prudence Rice thinks the choice had to do with an arbitrary reproduction of some more recent event in Maya history or with a culturally and historically significant date.[8] If it was an arbitrary reproduction, what dates are possible? Rice singles out the date 7.6.0.0.0 11 Ahau 8 Cumku (236 BC) as a possible candidate. For one thing, it falls right around the time when we find the earliest Maya Long Count inscriptions. And for another, it contains a whole number of *katuns*, an Ahau day name in the *tzolkin*, and a Cumku day name in the *haab*. Creation day corresponds with 13.0.0.0.0 in the Long Count and 4 Ahau 8 Cumku in the calendar round, so call it a triple bonus. Maybe the Maya back-calculated from 7.6.0.0.0 to a zero point that fit all those conditions. A second, less likely, possibility that has been mentioned is the date 6.19.19.0.0, which is 1 Ahau 3 Kej in the Olmec version of the calendar round, or 355 BC.[9] This Long Count date is just one tun short of 7.0.0.0.0; it too contains an Ahau creation day name and it corresponds to a winter solstice.

Because it is the sort of thing astronomers enjoy doing, I have also labored extensively over such calculations. I have never been able to find anything of cosmic significance, including the position of the Milky Way or the zodiac, that fits creation day. Maybe the date was only seasonally significant. The best I (and Rice) can come up with along that line is that the second of a pair of annual zenith passages—the August 11 date in the general zone of latitude where the Long Count may have originated—falls at a time in the seasonal cycle when people might have wished to come together to celebrate the completion of a successful crop. Not a bad time to re-crank your cycle.

As I said at the outset, deep-time reckoning is a widespread cultural phenomenon and it is often achieved via some sort of commensuration principle. For example, the starting point of the Julian calendar, fabricated in the sixteenth century and still in use by astronomers, is 4713 BC. It was arrived at by rolling back three different time cycles to a point of commensuration. One period included all possible combinations of the days of the week with the first day of the year, which amounts to twenty-eight years. The second cycle is the Metonic cycle of nineteen years, which tabulates the period over which a given phase of the moon comes back to the same date of the seasonal year. These two cycles are natural astronomical cycles, but the third is decidedly sociopolitical in nature— namely, the cycle of indiction, a period of fifteen years that originally marked the collection of taxes to be paid to troops discharged from the army; in other words, a monetary cycle. I cite this example to warn Maya calendar enthusiasts not to rule out the existence of time units that fall outside nature's realm. (Incidentally, the commensuration period of all three cycles in the Julian calendar [28 × 19 × 15] is 7,980 years; thus, in the course of this huge interval no two dates can be written down with identical entries in all three time cycles.)

Great cycles like the Julian era exist in calendars all over the world—Sanskrit, Hebrew, Chinese, and so on. I had always wondered why so many of them converge on zero points a handful of millennia BC. Could this be deep enough time for most civilizations to reckon in social terms? Our own modern culture is, of course, the exception, although I hasten to add that the vast stretches of time offered by Western science still perplex us.

To sum up, the lengthening of durational sequences in Maya timekeeping on the doorstep of the Classic period clearly must have been induced by a motive that drove a person or class of persons to propagate the notion that the present can be solidly anchored in the past by projecting events further back than anyone had hitherto contemplated. In the case of our own calendar, the prominence of the sole transcendent figure (Jesus) who lies at the foundation of the Holy Roman Empire takes front and center in the mission widely shared by complex cultures to extend deep time. In this case the Holy Roman emperor was the initiator of the great project of fabricating the architecture of time. I see no reason to think that the Maya were different, and I am convinced that the "bottom line" of the Long Count is directed not to the prediction of the cataclysmic end of time for all of us but rather to time's beginning and to the exaltation of the ruler, the one who initiated the Long Count project in the first place. That said about the Maya philosophy of time, we now turn to the role of astronomy and other natural events that are alleged to play a role in Maya creation 2012.

# THE ASTRONOMY BEHIND THE
# CURRENT MAYA CREATION

In Chapter 2, I noted the number of astronomical and other natural events conjured up by the Y12 prognosticators. To summarize, here is a short list of the most common questions I have been asked about possible natural events that might have something to do with December 21, 2012:

- Is there a unique galactic alignment that will take place in 2012?

- Could the Maya have known about it?

- Special or not, what effect can such an alignment have on the earth?

- Will a solar maximum occur in 2012?

- Might there be a cataclysmic effect on earth as a result?

- Will there be increased solar activity in 2012?

- Is there a connection between solar streams and unusual weather patterns on earth and could such a correlation produce cataclysmic effects?
- Is the earth's magnetic field weakening?
- Can this, perhaps together with any increased solar activity, wreak adverse effects on the earth?
- Is there a precedent for a magnetic pole reversal on earth and are we due for one?
- Might this have any cataclysmic terrestrial effects?
- Is the earth about to move into a hazardous region of the Milky Way Galaxy—some sort of "energy cloud" where shock waves or cosmic rays could cause disruptive effects on the sun?
- Could global warming, which some scientists say is produced by human-induced carbon emissions, actually come from the interaction of the sun, cosmic rays, high-energy particles from supernovae, and terrestrial clouds?
- Does the sun wobble and bulge because of the gravitational pull of the planets (or any other celestial bodies) appreciably enough to produce any sort of cataclysmic effect?
- Does the reversal of the sun's magnetic poles have any effect on the earth?
- Are the various natural disasters the earth is experiencing becoming more frequent than they were in the recent past as we approach 2012? Is the earth, then, experiencing a climax of sorts leading up to the year 2012?

In this chapter I will try, where possible, to offer answers to these questions based on the best available evidence.

So far we have established that August 11 (or 13), 3114 BC, the start of the current Maya creation, and December 21 (or 23), 2012, the date that marks its end and the start of a new creation, may have been deliberately set up to coincide with important times of the seasonal year. The beginning date of August 11 (13) is close to one of two annual dates when the sun passes overhead in the area where the Long Count calendar developed, and the end date coincides with the winter solstice, when the rising sun reaches its maximum southerly position as it swings like a pendulum annually along the horizon. In effect, the rhythm of the calendar starts out and/or terminates on a seasonal downbeat that may have imparted a sense of cosmic symmetry to the Maya system of timekeeping, much like the way we begin our day with both hands of the clock pointing straight up and once began our year with a solstice. There is supporting evidence for an early solstitial calendar in studies of the alignment of Maya architecture (see page 54). To judge from the batch of questions cited above, the big concern about 2012, at least in popular culture, is whether the early Maya calendar keepers were guided by events that operated on a longer time base than a seasonal one.

What do we really know about ancient Maya comprehension of cycles of natural phenomena? Given their unparalleled devotion to skywatching and mathematical computation, this question is certainly worth looking into. As we found in Chapter 2, one popular theory suggests that the 5,125-year Long Count creation cycle was geared to an alignment of the sun with the center of our Milky Way Galaxy, an alignment that occurs once every 26,000 years because of the precession of the equinoxes.

The galactic alignment hypothesis raises a number of questions about astronomy in general and about how astronomy was practiced in the Maya culture. First there are the astronomy questions: What is knowable about our Galaxy based on naked-eye observation? How precisely can such an alignment be established? What exactly is precession and how does it affect what we actually can observe in the sky? These questions are far easier to answer than the Maya questions: What did the Maya know about the sky—specifically, about the Milky Way? How did they conceive of the Milky Way (we already touched on this a bit earlier)? And what did they know about precession? We must answer these questions based on the evidence the Maya have left for us to ponder.

In my opinion there are few celestial thrills—a total eclipse of the sun and maybe a very bright aurora borealis, or northern lights—that can top a clear midsummer night's view of the Milky Way. Outdoors and away from city lights—a dark sky, an unobstructed horizon—the Milky Way appears as an irregular luminous band some two dozen full moons wide that arches across the sky. Slowly, the Milky Way changes its orientation throughout the night. As summer approaches in northern mid-latitudes, it can be seen encircling the horizon shortly after sunset. Then, as the earth rotates, it gradually arches upward, brightening. By midnight the starlit band crosses overhead, aligning northeast to southwest, about the way it is pictured in Figure 2b. Actually, the Milky Way covers 360 degrees of the sky. As Figure 12a shows, it appears as a nebulous ring all the way around the world because we are in the midst of it; therefore, we are unable to tell the extent of this great luminous forest because of the presence of the 200 million trees (stars) that light it up. Sandwiched more or less along the middle of the band is a meandering ribbon of dark interstellar matter—the stuff the stars are made of. This interstellar matter, which consists of dust and gas—the dust being responsible for most of the darkness—should not be confused with dark matter, dark energy, or black holes, all of which are dark for different reasons and invisible as well.

The dark ribbon, which looks like an absence of stars, roughly defines the plane of our Galaxy. It meanders its way among the constellations of the northern sky, from Cassiopeia through Cygnus and Aquila, where it becomes more prominent as the Great Rift. It widens as it approaches Scorpius and Sagittarius in the south. The discerning eye can detect the widening in the luminous band before it plunges below the southern horizon. The further south you go, the better the view, because the most luminous part of the Galaxy rises higher in the night sky.

The power of several centuries of collective astronomical observation and hindsight has revealed that the region of the Milky Way in Sagittarius, also called the "nuclear bulge," corresponds to the galactic center (the arrow in Figure 12a points to it); there the density of matter in the Galaxy peaks. Fortunately, the sun and its retinue of planets are positioned well out of galactic downtown—

we might say we live in galactic suburbia. Although this vantage point allows us to look toward the heart of the Galaxy, which lies some 25,000 light-years distant, we cannot see much more than one twentieth of the way to the nucleus because the interstellar dust, mostly carbon, absorbs the intervening light and thus blocks our view. If we could blow away the offending interstellar pollution with a gigantic fan, we would be able to read a newspaper by the light of the closely packed superluminous stars that surround the black hole that occupies the galactic center. That black hole is why I describe our location in the Galaxy as "fortunate." Galaxies, such as the one pictured in Figure 12b, are the building blocks of the visible universe. Our Milky Way Galaxy is but one of them. Billions of galaxies populate space as far as we can see.

In the West, the defining role of our own Milky Way Galaxy in the cosmic scheme of things only began to be realized in the late eighteenth century, when British astronomer Thomas Wright first suggested that the earth and the sun might be part of a gigantic wheel-shaped aggregate of stars. Wright's scheme came to be known as the "Grindstone Model" because he had likened the shape of the Milky Way to that then-common carpenter's tool. It was not until the early twentieth century that astronomers, equipped with far more penetrating telescopes than Wright ever could have imagined, were able to deduce the vast size and shape of the Milky Way Galaxy and our eccentric position within it. They did it by measuring the distances of bright variable stars known as Cepheids, whose periods of light variation are related to their absolute brightness. In other words, if you know the wattage of a light bulb, then how bright it appears to the eye serves as a measure of its distance. But even as late as the 1920s some astronomers positioned the solar system within a miniscule 300 light-years of the center of the Galaxy, which they perceived to be a slightly flattened aggregate of stars spanning a mere 4,000 light-years. (That is too small by a factor of twenty!)

One problem with basing alignment theories on the galactic center is that you cannot see it. Same goes for the plane of the Milky

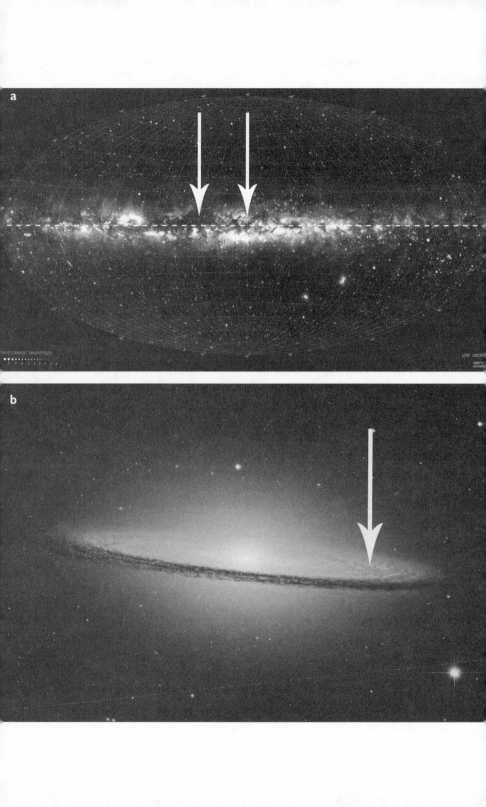

12. Galaxies inside and out: *a*. This 360-degree panorama of the Milky Way shows the position of the galactic center and the Great Rift to the left of it. The dashed line approximates the plane of the Milky Way. (Lund Observatory, http://www.noao.edu/swift/proposal/milkyway_lund_big.gif) *b*. The Sombrero galaxy is a system somewhat like our own Milky Way Galaxy. This advantageous view of it, along the plane of the "sombrero," enables us to see the dark, murky gap, which actually consists of the interstellar matter out of which the hundreds of billions of stars that light up the galaxy form. The arrow in the tip of the sombrero denotes where our solar system would be located if we lived in a neighborhood similarly positioned to our own but within this galaxy. (NASA/ESA and the Hubble Heritage Team STScI/AURA)

---

Way (the so-called galactic equator). Although it looks quite traceable on modern star maps (a danger I warned about earlier), the galactic equator was not even identified until the early 1950s, when astronomers were finally able to trace it with giant radio telescopes. Technically, the galactic equator is defined as the line that marks the highest density of neutral hydrogen emissions that emanate from transitions in the nucleus of that atom at a frequency of 1,420 megahertz (wavelength 21.2 cm). The galactic center is marked roughly by a prominent radio source known as Sagittarius A. Pinpointing the center of our Galaxy on the sky with the unaided eye by estimating the center of its widest point to a precision amounting to an area of about a dozen full moons might be possible. Put in golfing terms, this target width is comparable approximately to that of a fifteen-foot putt aimed at a hole about the width of a basketball hoop.

In sum, the Sagittarius region of the Milky Way is important in contemporary cosmology because we now know that it marks the center of our Galaxy, a vast system of matter and energy of which, in physical terms, our solar system is a tiny, insignificant part. The question is, why would marking the center of our Galaxy have been important to the Maya—important enough to have played a role in fixing their grandest time cycle? True, the luminous band of stars that makes up the Milky Way widens noticeably in the Sagittarius region, where the Great Rift that roughly defines the galactic plane is visibly enhanced. To the naked-eye observer, however, the region of the southern summer Milky Way does not appear to be the center of anything. They could have noted this widening in the stellar roadway, but the Maya record says nothing of it, although they did pay some attention to one of the crossing points of the Milky Way and the path of the sun marked by stars we call the zodiac.[1] Furthermore, as we learned in Chapter 3, Mesoamerican cultures in general conceived a universe consisting of layers, nine below us in the underworld and thirteen above in the heavens (see Figure 13). There is no evidence that they cared about spatial models with orbits and centers, nor even a spherical earth—all cosmological concepts inherited from the Greeks.[2]

What about the claim that the Maya observed the precession of the equinoxes? Precession is one of the longest astronomical cycles and, as we will learn in the next chapter, it is the fountainhead of so many theories based on cosmic determinism, ideas that attempt to link human destiny with the cosmos. That this 26,000-year cycle just happens to add up to five Long Count cycles (5 × 5,125.4 = 25,627 years) has added fuel to the hypothetical fire that, in their quest for cosmic harmony, the Maya may have stumbled upon one of the grandest of all celestial cycles. Because precession allegedly plays such a pivotal role in 2012 phenomena, we really need to probe its consequences in some detail.

Open an atlas to a map of the world with grid lines (latitude and longitude). Pick your favorite northern hemisphere town or city. Now imagine the grid lines slowly drifting. If they slide upward and to the right, then the latitude of the place you chose decreases, and its longitude increases. Take this exact situation, apply it to a map of the sky, and you have the nuts and bolts of precession. Star coordinates change.

Precession (from the Latin *praecessus*, or to "go before") is the slow marching along the ecliptic of the vernal equinox—the intersection of the ecliptic (the plane on the sky traversed by the sun against the constellations of the zodiac) and the celestial equator.

**13.** Mesoamerican cosmologies conceive of a layered, rather than point-centered, universe such as our own. This representation from early colonial central Mexico features nine underworld and thirteen upperworld layers. According to the Maya *Popol Vuh*, a dual-gendered creation deity presides over "all the sky earth." (Codex Vaticanus A, page 1, Akademische Druck-u. Verlag, Graz)

You can think of the latter as an extension of the plane of rotation of the earth onto the sky. Precession also shows up in the slow drifting motion of the place among the stars where the celestial pole (the extension of the earth's axis of rotation) is situated. This means that as the centuries go by we get a different north pole star—today, Polaris; 5,000 years ago, Thuban in the tail of Draco the Dragon; 10,000 years hence, bright Vega. The same goes for the south pole, where currently there happens to be no bright star (see Figure 14).

Precession produces a rather unsettling effect for long-term perspicacious skywatchers: things get out of joint. For one thing, the places on the horizon where stars rise and set change slowly through time (Figure 15). For another, the dates when stars make their first (or last) annual appearance in the sky after (or before) being lost from view in the sun's glare also change. So do the first days of all of our seasons. Thus, the sun at the spring equinox slowly migrates from one zodiacal constellation to the next. It was situated in the middle of Taurus when Egypt was in its prime, in Aries in the heyday of Babylon, and in Pisces at the time of Christ's birth. It will cross the border into Aquarius in about 700 years (see Figure 16a). This means that in 13,000 years the summer constellations (e.g., Lyra, Cygnus, Aquila) will change places with those of winter (Orion, Taurus, Gemini). There are a host of sound reasons to anticipate that any self-respecting skywatcher serious about his/her job might notice these changes.

The cause of precession is not really relevant here but, for those who might wonder, it results from the gravitational forces exerted by the sun and moon on the bulge produced around the earth's equator because of its relatively rapid rotation. To put it in animate terms, the sun and moon want the equator to line up with the ecliptic plane. The earth resists these forces by wobbling or gyrating like a top. One cycle of gyration takes 26,000 years. The full explanation of the cause of precession was not on the books until 1702, when Newton had devised the theory of gravitation. But even the great Sir Isaac had problems with the lunar force. That issue was

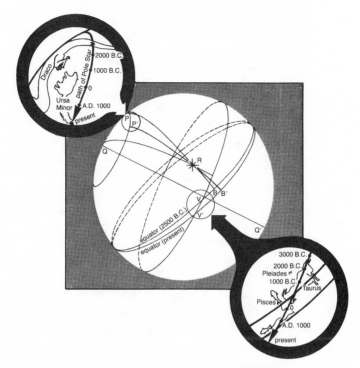

**14.** The precession of the equinoxes is defined as the motion of the vernal equinox along the ecliptic (from V to V' in this figure). As a result star R's coordinates change. Insets show an enlargement of the passage of the vernal equinox though different zodiacal constellations at different epochs (*left*) and the movement of the north celestial pole among the stars (*right*). (From A. Aveni, *Skywatchers: A Revised and Updated Version of Skywatchers of Ancient Mexico* [Austin: University of Texas Press, 2001], fig. 43)

not resolved until French mathematician Jean le Rond d'Alembert worked out the details in 1749.

The precession rate of the vernal equinox of 50.290966 seconds of arc per year leads to a full cycle in 25,770.1 years. (If you want to do the math, just take 50.290966 sec/yr × 1 min / 60 sec × 1 degree / 60 min, and divide all of that into 360 degrees.) Actually, the precession rate changes, slowly increasing with time as the angle between the earth's axis of rotation and the pole of the ecliptic

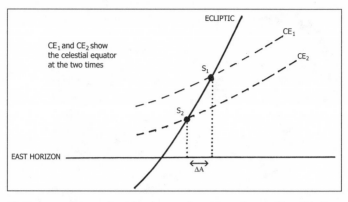

**15.** Because of precession, the change of position of a star ($S_1$ to $S_2$) along the ecliptic produces a shift ($\Delta A$) in the place where the star appears at the horizon. (Diagram by author)

(today approximately 23.5 degrees) decreases; therefore, in ancient Maya times the precession rate would have been less (e.g., if you use the precession rate effective 2,000 years ago, you end up with a full cycle of 25,053.2 years).[3] As we will see below, this affects some of the proposed interpretations of precession in the Maya record.

You can detect precession by measuring the difference between the length of the sidereal year (the time it takes the sun to return to the same place among the stars) and the tropical year (the time it takes the sun to get back to the vernal equinox)—about twenty minutes.[4] That difference can be measured by comparing the change in the distance of bright stars close to the ecliptic from well-defined points along the ecliptic, such as the vernal and autumnal equinoxes and the two solstice points. This is pretty much what the Greek astronomer Hipparchus did when he made his famous discovery of precession in 128 BC.

Hipparchus did it by spending years computing and tabulating the precise positions of 850 stars in celestial longitude and latitude (degrees along and perpendicular to the ecliptic) based on data recorded by his predecessors of the previous century or two. Hipparchus's ultimate goal was to create a star globe that plotted out the stellar positions accurately. (The Greeks were quite taken with making models—or simulacra, as they called them—that extolled the behavior of nature.) While constructing his star catalog Hipparchus noticed that the coordinates of the stars changed slowly with time. This had the unanticipated effect of changing the dates when the sun, moon, and planets entered and departed various constellations of the zodiac.

One way Hipparchus noticed the effects of precession was through changes in the simultaneous rising and setting of bright stars. Take Antares and Capella, for example. Where I live Antares disappears over the southwestern horizon just about the same time Capella rises in the northeast. But that was not true 200 years ago, when both stars were visible simultaneously (under ideal conditions, of course), and it will not be true 200 years in the future, when Antares will have left the scene in the southwest prior to Capella's appearance in the northeast. This is because the coordinates of both stars change. Precessional effects were quite bothersome to the ancient Greeks because their astrological predictions, the raison d'être for skywatching in the first place, could be altered drastically.

Hipparchus actually measured declination changes (i.e., a coordinate system based on the celestial equator rather than the ecliptic). He then used geometry to transform these observations to longitude differences. Unfortunately, this complicates things, because both the rates and directions of change in declination vary as a result of the 23.5-degree inclination of the celestial equator relative to the ecliptic. As Figure 14 shows, if you measure the change of position of stars in equatorial coordinates (right ascension and declination), you calculate a different rate of precession, depending on which stars are used and where they are located. In the example in Figure 14, the star R increases in both right ascension (from VB to V'B') and in declination (from BR to B'R') between a pair of epochs. On the side of the sky between ecliptic longitude 90 degrees and 270 degrees (basically the spring sky) declinations will decrease, whereas on the autumnal side, between 270 degrees and

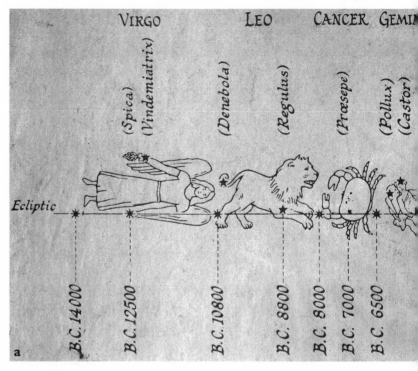

**16.** Are zodiacs universal? *a.* A segment of our Greek-Babylonian-derived Western zodiac. It consists of twelve constellations. The course of the vernal equinox sun through the zodiac during human history is marked out. *b.* The Maya zodiac in the Paris Codex (pp. 23–24), a pre-contact document likely dating to the fourteenth century, consists of thirteen constellations. Among those most easily recognizable are rattlesnake, bird, scorpion, and tortoise; they all hang from a serpent sky band. (Akademische Druck-u. Verlag, Graz)

90 degrees, they will increase. Furthermore, the biggest increases will be found near longitudes 0 degrees and 180 degrees, and the smallest will occur at 90 degrees and 270 degrees. So the best bet for calculating the precession cycle (once you are convinced it is a cycle) is to use data based on declination shifts of stars such as Spica, Regulus, and Arcturus on the spring side (0 degrees) and Aldebaran and the Pleiades on the fall side (180 degrees) of the

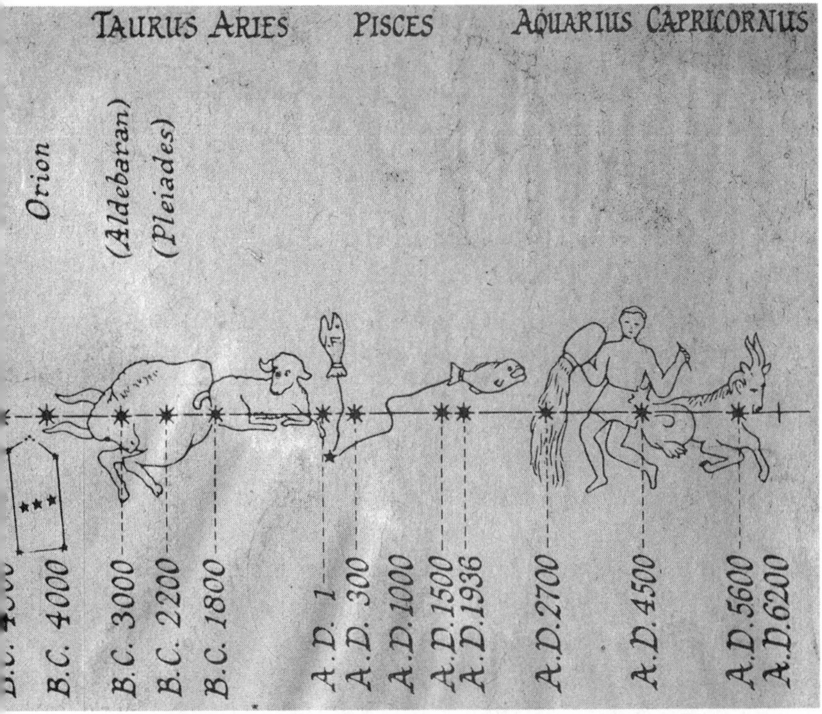

TAURUS ARIES PISCES AQUARIUS CAPRICORNUS

Orion

(Aldebaran)

(Pleiades)

B.C. 4000
B.C. 3000
B.C. 2200
B.C. 1800
A.D. 1
A.D. 300
A.D.1000
A.D.1500
A.D.1936
A.D.2700
A.D.4500
A.D.5600
A.D.6200

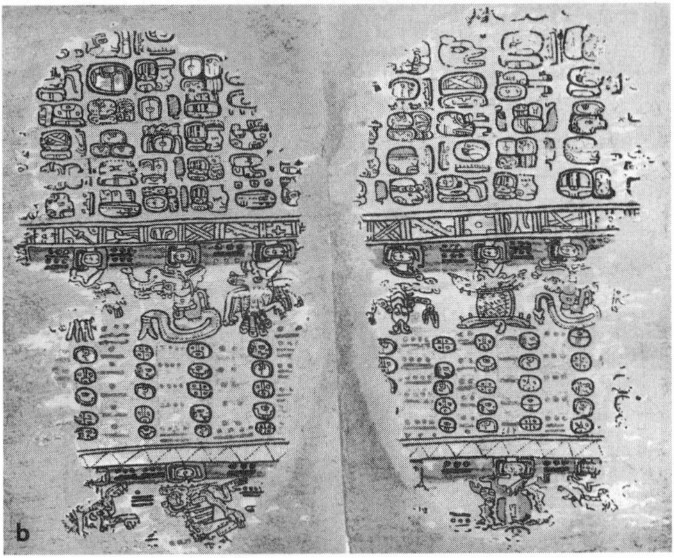

b

sky. (To add more difficulties, there are few bright stars close to the ecliptic on the autumnal side of sky.)

Hipparchus concluded that the drift of coordinates was common to all stars and that it amounted to one degree every eighty years; a full cycle would then take 28,800 years—not a bad approximation. It took more than 1,000 years of Greek and Islamic astronomical advances before the ninth-century Islamic astronomer Al Battani, thanks to lengthy tabulations of star positions based on a solar system modeled after a set of concentric spheres, improved the cycle to a more respectable 23,760 years, or one degree (about a day) of precessional shift every sixty-six years. Later Islamic astronomers called the motion a "trepidation of the equinoxes" (from the Latin *trepidus*, meaning "anxious"), because they thought it was based on two different conflicting motions. According to this theory, rather than slowly marching along the ecliptic, the equinoxes slowly oscillated about it. Finally, the early seventeenth-century Renaissance astronomer Tycho Brahe carefully measured the westward drift of the vernal equinox. He pegged it at fifty-one seconds of arc per year, which yields a precession cycle of 25,412 years, pretty close to the modern accepted value of 25,770.1 years, or a shift of one day every 70.56 years. Amazingly, all of this happened prior to the invention of the telescope.

Did other world cultures know about precession? The Chinese certainly did. A written record clearly attests to it. In AD 330 astronomer Yu Xi spoke of the *sui cha*, or "annual difference," a slight discrepancy between the length of the year measured by observations of the sun (the tropical, or seasonal, year) versus observations of the stars (the sidereal year), mentioned previously. There is no evidence, however, that Yu Xi recognized the *sui cha* as a continuous phenomenon; that is, he may not have known that it was cyclic.

What about the Maya and precession? We can be sure that the Maya had devised a zodiac, so surely they were concerned with the movement of the sun, moon, and planets along the ecliptic. The Maya zodiac (shown in Figure 16b) consisted of thirteen instead of our twelve constellations. But could they have detected the slow sliding of those constellations through the seasons? Possibly, but we have no evidence to suggest that the Maya ever developed coordinate systems based on the ecliptic or the equator, nor did they utilize the system of mathematical logic known to us as geometry. Moreover, the archaeological record offers us no evidence that the Maya possessed angle-measuring devices or sky globes. Among the few documents they have left us we find neither sky maps nor lists of tables of positions such as those that exist in the Old World.

But I do think there are places where we can turn to seek out data that might bear on the problem of precession's detectability. One might be in the area of astronomical alignments in Maya architecture. That the Maya oriented certain special structures to the horizon positions of celestial objects (especially the sun and Venus) is well-documented, although I hasten to add that we have no substantial evidence that they shifted their alignments to be in tune with changing celestial positions. There is one instance in Mesoamerica where, we have argued, that this may have been the case.[5] It concerns shifts in the alignment of the Pyramid of the Moon at Teotihuacan with the Pleiades. In addition, there is some evidence that the Maya may have utilized a base-20 system to mark off "degrees" of the sun's motion along the horizon.[6] Another place to seek out evidence on the Maya's knowledge of precession might lie in the study of the codices, which contain precise information the Maya gleaned from celestial observations. The Venus table in the Dresden Codex, for example, proves that the Maya were concerned with heliacal rise/set phenomena pertaining to that planet. With these documented skills at hand, how might the Maya have gone about measuring precession?

Table 1 gives the change in azimuth and in heliacal rise dates per century of several bright stars for the first half of the first millennium BC in the latitude of Yucatan.[7] The list includes some of the stars actually used by Hipparchus in his celebrated discovery of precession over two millennia ago (they are asterisked in the table). The first column of numbers gives the change in declination ($\Delta\delta$) per century toward the south (S) or north (N), taken from

Table 1. Coordinate changes due to precession (500 BC–0)

| Star | Δδ *(minutes of arc)*[*] | ΔA *(minutes of arc)* | Δt *(days)* |
|------|------|------|------|
| Antares | 75S | 30S | 1.6 |
| *Arcturus | 70S | 41S | 1.6 |
| *Regulus | 50S | 19S | 1.4 |
| *Spica | 66S | 35S | 1.4 |
| | | | |
| *Aldebaran | 75N | 28N | 1.4 |
| Belt of Orion | — | 20N | 1 |
| Capella | 46N | 28N | 1.4 |
| *Pleiades | 65N | 33N | 1.2 |
| Averages | 1.1 degrees | 0.5 degree | +1.4 days |

[*] After H. Thurston, *Early Astronomy* (New York: Springer, 1994), 151.

Hipparchus's original data (via Ptolemy in AD 150). The second column of numbers shows the shift in azimuth (ΔA) measured along the horizon in the latitude of Yucatan, and the last column gives the change in the date of heliacal rise per century.

Notice that all the azimuth shifts are smaller (and consequently more difficult to detect) than the declination shifts. This is because motion along the ecliptic (in longitude) when projected onto the horizon (owing to the relatively steep—and variable—angle of the ecliptic relative to the horizon) yields a smaller shift. The effect is clearly shown in Figure 15. Take, for example, the case of Spica, which lies close to the autumnal equinox. In Figure 15 the difference in longitude between positions S1 and S2 at two different times is approximately double the azimuth shift, ΔA, in the latitude of Yucatan.

Clearly, if they used these data the Maya *could* have detected the effects of precession. After all, 0.5 degree (thirty minutes of arc) is the size of the moon's disk. That much change over a century is quite noticeable. Whether the ancient Maya actually perceived the cosmic shift as a constant and continuous effect (an issue debated in the West for more than 1,000 years after Hippachus) is quite another issue. But for now we will be optimistic and suppose that they did.

If we calculated that the average shift in star azimuths is 0.5 degree per century, as demonstrated in the table, then how many years would it take for the star positions to drift all the way through the seasons and get back to the same point? The answer is 72,000 years. This is quite different from the result obtained from a sample of Hipparchus's declination-based data, where a 1.1 degree shift per century yields a precession period of approximately 32,700 years (and recall that Hipparchus's best attempt landed him much closer to the accepted answer). So we must conclude that an *accurate* Maya precession cycle is not to be found in data based on building horizon alignments.

Let us turn next to our heliacal rise data (in the righthand column of Table 1). With date shifts averaging 1.4 days per century, our answer comes out to a quite respectable 25,700 years for the precession cycle. But let me add two cautionary notes. First, that spot-on result of 25,700 years is a bit misleading because there is a 2–3 day error (we will call it 2.5 days) in accuracy in fixing helia-cal rise dates. So, an advance of seven days over a 500-year baseline amounts to a little over one-third tolerance in our calculations. Consequently, it is more correct to say that the Maya precession period, if determined from heliacal data, would amount to some-thing between 17,000 and 34,000 years. Second, our discussion of Hipparchus's determination of precession is based on hard data tabulated by him and later by Ptolemy in star catalogs. In other words, it is based on evidence obtained from the cultural record. On the other hand, our analysis of the Maya side of the question of precession's detectability is entirely hypothetical, relying on our perception of sky changes Maya astronomers *could* have observed.

For all of these reasons I am forced to the conclusion that the Maya certainly could have detected precession (which I have sug-gested before[8]), especially from heliacal data, and that they could have approximated the precession cycle to within a few thousand years. Whether they did so requires historical/archaeological evi-dence. Let us turn, then, to an examination of some of the claims that the Maya actually measured precession.

Because they contain so many long intervals, the inscriptions offer excellent fodder for those attracted by lengthy astronomical periods. One concerning the Maya discovery of precession in the epigraphic record[9] highlights an 8,660-day (1.4.1.0 in Maya) interval that appears on a carved bone from a royal burial at Tikal. It is accompanied by three phrases that read "first 11-pik," "second 11-pik," and "third 11-pik." The accompanying three calendar round dates, translated into Long Count, are all separated by 1.4.1.0. They also mark the first three inscribed dates separated by that interval from 13.0.0.0.0, the creation date. An interval of 11 *baktuns* also leads to the same successive calendar round dates. This set of intervals may have functioned as a time cycle to track dynastic events at Tikal and other sites relative to the creation episode. Three *pik* cycles add up to 71.18 years, a fairly long life for any Maya person.

You may already have guessed where precession comes in. Since the sun moves one degree per day along the ecliptic on its annual course, the one degree of precession (mentioned above) measures the time it takes the solar (365-day) and sidereal (366-day) years to misalign by one day. Three *pik* cycles happen to fall close to one day's worth of precession. How would this sort of shift be detectable? As discussed above, Maya astronomers could have used heliacal rise/set dates, which are documented in the codices, although not in the context of anything having to do with precession. The changes are large enough to be detectable (up to a few days per century) but, as we have seen, they are also highly variable. Or you could track "one day of precession" by noting the slow shifting of stars in the zenith, or the shifting dates of solar relative to stellar zenith passages, as some investigators have suggested.[10] Once again, however, the problem is that stars shift at a variable rate. You could average together rates of azimuth shift and heliacal rise and set dates among a large number of stars. Question is, can one mount an argument to show that the Maya actually did this?

In his study of the large numbers on the "Serpent Pages" (pp. 61–69) in the Dresden Codex, Mayanist Michael Grofe argues that the Maya computed the precession period.[11] He reads the introductory number to these pages as a whole number of sidereal years. He places another number, the so-called Serpent Base Date, written in calendar round notation, on the summer solstice more than 30,000 years before the Maya creation date in 3114 BC. Then he notes that the length of the seasonal year (established from Copan inscriptions) drifts by some 215–218 days over that time frame. But that is equal to the anticipated error difference between the three-*pik* 71.18-year period and "one day of precession," which we today tabulate as 70.56 years. So, he concludes, the Maya were picking up the drift.

One problem with this theory is that Grofe uses the *contemporary* value of the rate of precession to mount his argument. But as we showed earlier, the rate of precession varies. In the first few centuries BC, when the Maya devised the Long Count, one day of precession was equivalent to 69.96 years. Using observations that could have been made then—not now—Grofe's 215–218 day shift really amounts to 446 days.

There is not space enough here to fully critique Grofe's work, which has acquired uncritical acceptance in some quarters. Still, a few important points need to be made. First, on the epigraphic side, Harvey and Victoria Bricker, who have recently concluded a detailed study of astronomy in the codices, suggest that Grofe has misread many of the key large numbers as well as the glyphs he uses to apply these numbers in the Serpent Pages to his astronomical interpretation.[12] Second, regarding projections back 30,000 years, we do not know enough about the variability of astronomical periodicities to project sky views back confidently to much more than a few thousand years BC. Anyone who cherry-picks big numbers from diverse sources is bound to discover whole multiples of diverse astronomical periodicities.

Based on the way the Maya bare their enthusiasm about predicting eclipses or timing the aspects of Venus and Mars, my own view is that if the Maya really cared about precession, their way of showing it would likely consist of combining recognizable short periods into bigger ones that are commensurate. When it comes to precession the obvious pair would be 365 and 366 days, the lengths of the solar and sidereal years. If precession were such a big deal to them, I would anticipate finding something like a 365/366-day table,[13] with a correction page in the codices. One problem with all Maya precession theorists is that they offer no mechanism extant in the Maya record. At this stage, I think it is likely that the Maya knew that what we call "precession" existed, but to date there is no evidence to support the case that they calculated the cycle, much less even perceived precession as a cyclic phenomenon.

Finally, and just for the sake of argument, suppose the Maya did calculate precession and that they marked the Milky Way's plane and the galactic center as well. How could they go about detecting when the sun at one of its positions in the precessional cycle—for example, the December solstice-ending date of the contemporary era—would pass through the galactic plane, as Jenkins suggests? Since you cannot see the Milky Way when the sun is out, you would need to approximate its location by noting dates (and remember that all of them vary) when guide stars in adjacent areas first appear or disappear during twilight; then you would need to perform some sort of interpolation to get a fix.

Same goes for the galactic center. All things considered, the best estimate I can come up with for the tolerance in timing the crossing point of the solstice sun and the heart of the Galaxy is at least a hundred years.[14] For the naked-eye observer, then, the winter solstice sun has been crossing the galactic center every year since the late nineteenth century and it will continue to do so until the early twenty-second century. I doubt very much that the Maya would have deliberately devised and targeted an end to the Long Count on December 21 (or 23), 2012, based on these observational data.

Before we exit this lengthy, but necessary, discussion, it is worth pointing out that despite the lack of evidence concerning the Maya record and precession, there are plenty of reasons for admiring the Maya astronomers, who we know were quite capable of charting astronomical phenomena. For example, studies of the Venus Table in the Dresden Codex, a long-term predictor of first and last appearances of the planet Venus in the sky, prove that the Maya were well aware of when to apply small corrections to their calculations to keep their almanac on course with the planet. This is why we often see it written that the Maya could time the heliacal rising of Venus to one day in 500 years (I have said so several times myself!) or that they calculated the period of the phases of the moon to an accuracy that rivals that of modern astronomy. Because they cannot imagine how the Maya could have accomplished these feats without high-tech equipment (such as telescopes and computers), some enthusiasts endow the Maya with supernatural senses and powers—and that is where lost Atlantis and extraterrestrials often enter the picture.

In fact, the Maya scaled these astronomical heights not because they had super brains or super equipment that yet may lie buried among the ruins, but rather because they were super *persistent*. Take the phases of the moon as an example. Suppose you and I note the date of the next full moon and then count the number of days to the full moon that follows it. This is not easy to do because full-moon dates are difficult to nail down; the moon could be slightly deviated from perfect roundness on either side when one or the other of us makes the call. Say you estimate twenty-nine days and I come up with thirty. No matter. Let us count days to the next full moon and the next and the next, and so on. Suppose that we, like the Maya who devised their eclipse warning table in the Dresden Codex, keep that record of counting going for 11,958 days, give or take a day (approximately thirty-two years, which corresponds with the length of the Eclipse Table). Despite our short-term disagreements, in the long run we would both tabulate 405 full moons over that long interval. And we would both arrive at an average length of the lunar month of 29.525925 days. Compared with modern astronomy's figure of 29.530589 days, that is a difference of only seven minutes. The same kind of methodology applies to Venus watching or eclipse predicting. Naked-eye observations of short-term periodic phenomena like these, taken over long periods of time—phenomena quite different from precession—can be time averaged very precisely. And that is exactly what the Maya did. What is most amazing about Maya astronomical achievement is the attention span of the Maya astronomer.

So much for the stars. Now, what are we to make of claims about Maya knowledge of other natural phenomena, such as the sunspot cycle and the reversal of polarity of the earth's magnetic field? Are there coincidences that point to a Y12 cataclysm? As with precession and the Milky Way, we need to look at these phenomena, how they operate, and what we can say about them with some certainty.

The sun is a dynamo. In its 10 million–degree interior furnace, electrically charged gas particles churn about in a pair of conveyer belt–like cells, one in the northern and the other in the southern hemisphere. The moving charge creates a magnetic field that fluctuates in intensity. All active phenomena on the sun—sunspots, flares, prominences—gradually build in strength and number as the field intensifies. The phenomena peak, and then wane as the field reverses—south pole becomes north and north becomes south. Then the process happens all over again on a time cycle of eleven years (or twenty-two years if you include the cycle of polarity reversal).

Sunspots are cooler areas located just above the photosphere, or visible surface of the sun. When Galileo first glimpsed them in 1610 through his hand-held telescope, he thought sunspots were holes in the sun's surface. Under extraordinary circumstances, say sunset or sunrise on a clear day, they can be glimpsed on the solar surface by a perspicacious naked-eye observer, but prior to the telescope the sunspot cycle went undetected. Sunspots are caused by the explosive issue of gas just below the photosphere. The magnetic field conducts the hot gas to heights of thousands of miles where it cools. So the spots, at temperatures averaging 5000°K, appear as dark blotches against the background of the 6000°K photosphere. At the beginning of an eleven-year cycle sunspots are small and relatively few in number. They appear at mid-latitudes on the solar surface. As the cycle intensifies, more and more—and generally bigger and bigger—spots develop. Many are larger than the earth. The biggest one on record is the Great Sunspot of 1947. It was more than forty times the size of the earth. Toward the end of a cycle spots begin to crowd the equatorial regions of the sun, disappearing just as a fresh crop breaks out at the mid-latitudes.

Sunspot numbers have been charted since the invention of the telescope. Between 1645 and 1715 there were rather few of them. This was the so-called Maunder minimum, named after the astronomer who first pointed it out. Interestingly, this period coincided with the Little Ice Age, an extended period of very cold winters in Europe. That coincidence, I think, is largely responsible for the popular notion that sunspots cause drastic climate change. Actually, there were three frigid minima centered around 1650, 1770, and 1850, with warmer periods between. To date, no physical connection between the Little Ice Age and sunspot activity has ever been established. Yet, although climatologists have no generally accepted theory to explain it, there are some correlations between weather and sunspots. One research group, for example, has noted that unusually heavy rainfall in East Africa tends to occur about one year before a sunspot peak, but little data exists to reach any solid conclusions about the underlying cause of this correlation, if it is real.

At the opposite end of the spectrum, 1958 was a banner year for sunspots, with a whopping number of 210 recorded at peak. The peaks of activity seem to fluctuate over long periods, although a super-period for sunspot numbers is hard to pin down. Since the 1958 maximum (cycle number 19) no cycle has approached 200. Cycle 20 in 1969 was shallow and broad, topping out at about 130; cycles 21 in 1980 and 22 in 1990–91 produced sunspot numbers around 180, whereas cycle 23 in 2001–02 was a virtual replay of cycle 20. Next up on the solar schedule is cycle 24, which is due to top out in 2012–13. Predictions point to a peak somewhere between 150 and 180—hefty, but nothing unusual. How strong all of this stormy activity at the peak of any cycle will be in the future depends on how fast the conveyer belts in the solar interior turn. A recent slowdown of record proportions portends a drop in activity about two cycles in advance. That means that although cycle 24 will be moderate to heavy, cycle 25 (scheduled to peak in 2022–23) could be one of the weakest.

Y12-ologists who posit that the Maya might have been aware of some sort of impending disaster on our temporal horizon will not find a whole lot of support in either the solar or the Maya record. There is no eleven-year cycle, or multiple thereof, cited in Maya records; nor is there any evidence that the Maya saw or cared about sunspots. The same holds for solar flares, which are the opposite of sunspots. These superhot areas above the solar photosphere reach temperatures of 7000°F and usually develop above sunspots. With higher temperature comes more high-frequency radiation, and consequently a heavier dose of ultraviolet radiation, from which we are fortunately (at least for now) protected by the ozone layer in the earth's atmosphere. (We can take further comfort from the fact that we here on earth receive only one two-billionth of whatever energy blasts out of the sun.) Once a flare erupts, a stream of high-energy particles (or protons) departs the sun, reaching the earth in a day or two. When it gets here, the earth's magnetic field draws the protons toward the magnetic poles. When they enter the upper atmosphere they collide with oxygen, nitrogen, and other molecules, transferring their energy to them. As the molecules de-excite they emit the beautiful reds, greens, and yellows that make up the aurora borealis, or northern lights, one of the most impressive celestial spectacles you can witness (as I opined earlier). We take particular notice when flares cause radio-TV fadeouts by making the ionosphere more transparent so that earth-based signals cannot bounce back toward the ground. (I do not think the Maya had radios.)

Aurorae are rare happenings in tropical latitudes. They occur with greater intensity closer to the magnetic poles but some have been reported in the tropics. I recall receiving an anxious call from a fellow Mayanist who had witnessed an auroral display at Tikal during the 1980 solar maximum. He was *very* excited. Maya skywatchers surely would have known of the colorful lights in the sky, but if they expressed what they saw, I have yet to witness it.

Over the long course of observing flares, solar astronomers have noted exceptional ones. In the summer of 1956, for example, the grandest flare in history occurred, a rare event that produced a 2 percent increase in the cosmic-ray flux incident on the earth's atmosphere. Fortunately, this fluctuation was not enough to do any damage. In the past two decades we have heard more about them; NASA and other Web sites monitor flares, displaying dazzling pictures. I wonder whether Y12 doomsayers tend to exaggerate the intensity of flares seen in recent times simply because we have the space technology with which to observe them more closely and we now know that they have a noticeable effect on earth. In June 2005, for example, a strong storm of high-speed protons zapped computer circuitry and messed up satellite communications. Any astronaut performing out-of-vehicle operations in space at that time could have suffered significant radiation damage.

Add to the Y12 celestial scare list cosmic rays. They seem to worry us even more than sunspots and flares. Cosmic rays are high-energy particles that enter the earth's atmosphere from the neighborhood of the solar system. In large doses, such as might be produced if a nearby star became a supernova, they can cause damage, such as an enhanced risk of cancer or cataracts. Once again, the earth's atmosphere comes to the rescue, diminishing their effect—with an assist from an unlikely quarter. Ironically, the solar storms I mentioned earlier blow away cosmic rays. In effect, we trade one celestial evil for another. So much for getting zapped by weird radiations. Barring the occurrence of a sudden nearby stellar explosion or an unanticipated overturn in the sun's interior—the odds of either are incalculably small—there is nothing to fear under the sun.

As we learned in Chapter 2, galactic forces emerge as a major focus among the prophets of Maya creation. They tell of black holes at the galactic center that will change the world, or they stress the urgency to reconnect with our cosmic heart through galactic alignments. Is earth affected by galactic or other cosmic forces? According to one popular theory the earth periodically enters different areas of the Galaxy, where cosmic rays and other harmful activity threaten life's existence. True, the solar system does oscillate back and forth above and below the plane of the Galaxy as it revolves every 240,000 years about the galactic center, periodically passing through the spiral arms where most of the traffic in the firmament flows. Astronomers still debate possible effects from the periodic passage of the solar system through the spiral arms of our Galaxy. Ten million years ago, for example, the solar system emerged from the Orion Spiral Arm. There we encountered more interstellar matter than normal, but when you realize that its density (about one atom to the cubic meter) is several orders of magnitude below that of a laboratory vacuum, it is difficult to imagine that any perceptible effect, such as the advent of an ice age (which has been suggested), might result.

There is now little doubt that a major extinction event at least played a role in wiping out the dinosaurs some 65 million years ago. An asteroid or comet nucleus about five miles across may have been responsible. Periodic mass extinctions of species also may have occurred in the distant past. From the study of fossil remains, paleontologists back in the 1980s thought they found extinction episodes that followed an approximately 26-million-year cycle over the past 225 million years.[15] Since then closer studies have placed these findings in doubt. The periodicity, if it is real, has no acceptable explanation to date. If a dark nearby planet or star is responsible for such catastrophic effects, it would have been easily detected because of its gravitational pull on the outer planets, causing deviations in their orbits. So we do not really have any concrete evidence about mysterious galactic forces affecting the earth. There the matter rests.

What about forces emanating from magnetic fields, another popular Y12 claim? The earth, of course, has its own magnetic field, but it is only one-third as strong as the sun's. Credit the ancient Chinese with the invention of the magnetic compass (sometime before AD 1000) and the lodestone, one of the iron oxides with the capacity to detect magnetic forces. Magnetite has been found in Maya offertory caches, but there is no solid evidence the Maya ever used it to make a compass.[16] We know that the magnetic field of the earth fluctuates slightly on a daily and yearly basis, as well as over long periods of time. Studies of paleomagnetism based on measuring fossilized magnetic field directions and strengths in ancient fire pits tell us that the earth's field overturns just like the sun's. It exchanges north and south poles, but on a much longer time scale. Recent magnetic field maps show an accelerating movement of the north magnetic pole. Polar flips happen about every few hundred thousand years on average. The last one occurred 780,000 years ago and the whole process took several thousand years, so precise predictions are clearly out of the question.

What is in store for us when a field reversal happens? Although weak, the field is nonetheless strong enough to affect the way some animals navigate and it has played a major role in helping humans find their way around too. Communications might be affected. You may even see multiple auroras over several poles as the field gets more complicated. Beyond that we do not know much; but since the process takes so long, polar reversals can hardly be described as "cataclysmic."

What about gravitational effects? Do they portend major changes in our environment anytime soon, as some have suggested? As the earth and the other planets circle the sun they tug at it from side to side. In the solar system, gravity is basically a game of seesaw. In equilibrium the more massive object lies closer to the center of balance, or fulcrum, the less massive farther away. If, for example, a 150-pound adult and a 50-pound child hop on a seesaw, the child needs to be three times farther (150 ÷ 50) from the fulcrum to achieve balance. Now the sun is fifty times more massive than all of the planets combined, so none of them, with the possible exception of Jupiter and Saturn, which together possess 98 percent of the mass of the planets, makes much of a difference when it comes to jerking the sun around gravitationally. Let us suppose Jupiter and Saturn act in unison; that is, they line up on the same side of the sun (which, incidentally, they do every couple of decades). Then the center of balance will be approximately one-fiftieth of the way between where they are situated, or roughly 13 million miles from the center of the sun (about ten sun diameters). This effect has been taking place for 4.6 billion years without doing much damage to the solar system.

Solar system bodies also raise tides on one another by pulling more strongly on the side facing them. The strength of the tide depends on the mass of the tide-raising body and how far away it is. Tide-raising forces on the earth are produced principally by the moon (which is very close to the earth) and by the sun (which, although farther away, is much more massive). The moon produces a two-foot tide and the sun a one-foot tide in the ocean.[17] When they line up (i.e., when there is a new moon or a full moon), the two tides enforce one another, creating a three-foot tide. When they oppose (i.e., when the moon is at right angles to the sun as seen from the earth), which happens at first and last quarter moon, the ocean tide on average is only one foot high.

Even if all the planets in the solar system line up with the sun—and the earth—the difference in the tide-raising force compared to what we normally experience on average is not enough to change what is happening in the interior of the sun—or the earth. The effect would be about equal to that of a few dozen sumo wrestlers attempting to topple the Empire State Building by lining up on the same side and pushing.

---

At the beginning of this chapter, I culled from the literature, both on and off the Internet, what I consider to be the most frequently asked questions about natural phenomena in relation to predictions about 2012 and the Maya. Then I tried to offer the best answer that I could to each of the questions based on the evidence in the possession of those who specialize in the study of these phenomena. Given the litany of potential Y12 cosmic threats I have tried to address, even if only briefly in this chapter, let me sum up my responses to these oft-encountered geological, geophysical, and astronomical questions.

Concerning any unique galactic alignment that will take place in 2012, we can say that over a period of a few hundred years on either side of the year 2000, the sun at winter solstice indeed will cross the plane of the Milky Way Galaxy. Whether the Maya knew about it is another question. I have shown that it is possible, but also that there is no solid evidence to verify it. Even if they did, I know of no effect such an alignment might have on the earth. I have also gone to great lengths to demonstrate that although Maya knowledge of precession of the equinoxes, implied by the galactic alignment theory, is possible, no one has successfully demonstrated that the Maya calculated that cycle.

Most of the other claims about Maya awareness of cataclysmic natural phenomena are presented in the Y12 literature without reference to any data whatsoever in the Maya record. Nevertheless, because of the scare literature connected with them, I felt it necessary to explore the likelihood of their affecting the earth. Concerning solar activity, yes, there will be a solar maximum in 2012. Based on examining all the data I do not understand how a cataclysmic effect on earth might result. Although there will be increased solar activity in 2012, the peak of the eleven-year cycle, by no means will this activity be record-breaking.

Although nothing in the Maya record points to an awareness of these phenomena, I think solar streams and unusual weather patterns on earth can be weakly correlated. But a correlation does not necessarily imply a cause. In any event there are no effects here that I would describe as potentially cataclysmic. It has been suggested that the weakening of the earth's magnetic field, together with increased solar activity, could wreak serious and adverse effects on the earth. True, the field is weakening, but based on what scientists know about solar-terrestrial effects, to describe these effects as "serious and adverse" would be too strong a statement. Earth is in the incipient stages of a long-term terrestrial polar-field reversal. It should happen over the next several thousand years and the process itself will take a few thousand years—another long-term effect that cannot be pinned down to 2012. On whether such a reversal can have any adverse terrestrial effects, we simply do not know enough to give a definitive answer. Communications certainly will be affected—and perhaps climate change and animal behavior as well. But again I must stress that the process is *gradual,* so the word "cataclysmic" seems too strong a term to me. The sun also reverses its field, and that polar reversal does not affect the number of charged particles streaming out of the sun at that point in its cycle.

Concerning the possibility of earth-rending forces of greater magnitude, we know that the solar system oscillates periodically through the spiral arms in the galactic plane, but I know of no phenomena involving larger scale forces (e.g., "energy clouds" where shock waves or cosmic rays could cause disruptive effects on the sun) that imply that the earth is about to move into a hazardous region of the Milky Way Galaxy.

We know that the sun does wobble and bulge because of the gravitational pull of the planets, but again the effect is too small to produce any sort of cataclysmic effect. Global warming, which most scientists agree is produced at least in part by human-induced carbon emissions, has no demonstrable connection with any of the aforementioned forces. And although the Maya may have been affected by climate change, especially during the collapse period, I know of no evidence that they recorded climate cycles.

Finally, natural disasters on earth have become neither more nor less frequent as we approach 2012 than they were in the recent past, although those who cherry-pick scientific data (a hurricane here, an earthquake or sunspot there) might think otherwise. So the earth is *not* experiencing a climax of disastrous events pointing to the year 2012.

Why then all the fuss? To really appreciate and understand the 2012 phenomenon we need to place it in a broader historical and cultural context. In the next chapter, I will carry the fascinating story about predicting the end of time well beyond the realm of the Maya. There we will discover that Ecclesiastes 1:9–14 may have had it right: "What has been will be again; what has been done will be done again; there *really* is nothing new under the sun" (my italics).

# WHAT GOES AROUND:
# OTHER ENDS OF TIME

The Maya were not unique in their creation of ever larger time cycles that transcend seasonal years; for example, we reckon the ten years of a decade, the hundred of a century, the thousand of a millennium, each seeming to take on a character of its own. The Maya thought of *katuns* as we think of our decades, as a way of labeling patterns of social behavior. In America, the 1890s were gay, the 1930s depressing, and the 1960s revolutionary and marked by idealism, yet marred by assassinations. On a larger time scale we speak of the Dark Ages, the Middle Ages, the Age of Enlightenment, the Industrial Age, and so on. Then there is deep time.

Eras are even longer than ages. We think of a chronological era as a succession of years that proceeds from one fixed point in time to another, often commenced by a seminal event. The birth of Christ, for example, gave rise to the Christian era. For the staunch believer, the end of this era will arrive with the second coming of Christ, an idea bolstered in the early Christian mind by the belief that all things happen expressly for the ends that they fulfill. Time pulls us forward into the future. For the Christian, the end of the old era will culminate in the arrival of the kingdom of God, which will initiate a new era—a timeless eternal existence to be experienced only by the devout believer.

Prior to the early Christian era the sense of long duration in the West had been rhythmic and repetitive. The Greeks marked the Great Year, the period of creation of the world to destruction and rebirth, by the time it took the sun, moon, and planets to get back to the same positions they occupied in the previous round of time. For the pessimistic Greeks the five ages of creation followed a downward-spiral course designated by the degree of decline in value of the precious metals after which they were named. First, a too-perfect *golden* race of mortals lived like gods. In an abundant land free of labor and misery, they never grew old. After them the gods created a second, *silver* race where lives were shorter and more troubled. Because these people refused to worship the gods, the latter were banished to the lower world. A third *bronze* race, a warlike people who were grotesquely deformed, replaced the silver race. A plague was sent to remove them from the face of the earth. In the heroic age, the fourth race of demigods that preceded us was given a boundless earth, but they too were destroyed, this time by the evils of warfare. "We live in the age of the iron race," wrote the ninth-century BC Greek poet Hesiod.[1] This is why we toil by day and anguish by night. We will know when Zeus will do away with our race, for our children will be born old and there will be great tension between parent and child. "Would that I now were no longer alive in the fifth age of men," laments the poet.[2]

As I suggested in the last chapter, Great Year cycles with astronomical underpinnings, like the Julian era, are present in calendars from civilizations the world over. The Indian calendar, adopted from the earlier Sanskrit model, tallies a 2,850-year Great Year made up of 150 Metonic cycles. Theirs was a calendar composed of mathematically precise relationships and socially significant time intervals. It has a celestial zero point—a hypothetical conjunction of all of the visible planets in the constellation of Aries, an event they calculated to have occurred at midnight on the night of February 17–18, 3100 BC.

Oddly enough, the theory of the flood in Greek astrology uses the same conjunction date. The Greek zodiac, which charts the course of the sun, moon, and planets among the stars, is divided into four "triplicities," each tied to one of the four elemental qualities:

> Aries, Leo, and Sagittarius are fiery.
> Taurus, Virgo, and Capricornus are earthy.
> Gemini, Libra, and Aquarius are airy.
> Cancer, Scorpius, and Pisces are watery.

Close encounters of the planets seen in the sky in any of the first three signs portend destruction by conflagration (the multiple conjunction of 3100 BC happened in this triplicity), whereas a conjunction in any of the last group presages destruction by flood. Shifts between triplicities also warned of major dynastic changes, and a completed cycle meant the arrival of a major prophet. Chinese astrology also lent great importance to close gatherings of the planets. One theory holds that the zero point of the Chinese calendar also was based on a back-calculated multiple planetary conjunction.

Cosmic re-creation cycles, also called world ages, operate in frameworks of different length among different cultures of the world, but the celestial associations that guide their underlying structure is pretty much the same. The Chaldeans, for example, wrote that the universe would be deluged when the seven planets were assembled in Cancer, and that destruction would proceed by fire when they arrived in Capricorn.

The Inca of Peru attributed their creation to a male-female deity named Viracocha, who emerged from the waters of Lake Titicaca (in Bolivia). Because the *khipus*, knotted string devices on which Andean people kept their records, have yet to be fully deciphered, we have no handle on the periodicities involved. Spanish chroniclers tell us, however, that there were four world ages (Figure 17a).

Viracocha's first creation was a race of giants who stumbled about in the dark. Dissatisfied with his own work, he flooded the world and turned the giants into stone. (You can still see them today in the mountains that surround Cuzco, the Inca capital.) Then Viracocha began a second creation. He called forth the sun, moon, and stars from out of the great lake (Lake Titicaca) to light the world, and he began time by setting all of them in motion. In a later age Viracocha made humans, modeling them in the clay that he found at the shore of the lake. He painted each of them with a colorful dress so that he could define their kinship group, and he endowed them with speech and song and dance and gave them food to eat.

Another common denominator in the narratives of many cyclic creation stories is that with each cycle or age the creators improve on their work. As we discovered earlier, the Greek world ages are an exception because they deteriorate as time marches on, going from gold to silver to bronze, and then after the heroic interlude, iron. The famous Aztec Sun Stone, pictured in Figure 17b, is a pictorial narrative of a similar cyclic creation story. Its inner panels show the pictographic calendar signs of the means of destruction of the four previous creation eras, or "suns." The first was the Sun of Jaguars, which ended when a population of giants was eaten by ocelots or jaguars. Then came the Sun of Wind, which ended in hurricanes, followed by the Sun of Rain, when flooding did in the world, and finally the Sun of Fire, when a great conflagration destroyed the world. We live in the fifth sun, the Sun of Earthquake, or Movement, which foretells the means of our demise—unless we act to avert it by providing Tonatiuh, the god with the lolling tongue pictured at the center of the stone, with the debt payment of the blood of sacrifice.

Anyone who experienced Hurricane Katrina in 2005, the Great Chicago Fire in 1871, the periodic devastating floods of the Mississippi River, the great Indonesian tsunami of 1883, or the 1906 San Francisco earthquake understands why cataclysmic endings play a major role in the stories of so many civilizations. These are among the most life-threatening natural occurrences humanity

**17a.** Other end-of-the-world accounts: The Four Ages of the World, according to the Inca chronicler Felipe Guamán Poma de Ayala, is a revised, post-contact version of an earlier story. It pictures the progress of civilization from an age when peasants clad in forest garb tilled the land (*top left*) to one in which their descendants were forced to live in an uncertain world among their conquerors. (J. Murra and R. Adorno, eds., *El Primer Corónica y Buen Gobierno* [Mexico City: Siglo XXI Editores, 1980], used by permission of Siglo XXI Editores, Mexico City, from the first unabridged edition of the seventeenth-century manuscript)

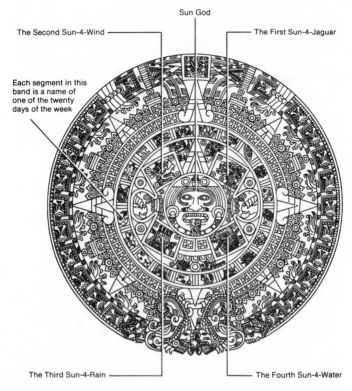

Sun God

The Second Sun-4-Wind — The First Sun-4-Jaguar

Each segment in this band is a name of one of the twenty days of the week

The Third Sun-4-Rain — The Fourth Sun-4-Water

**17b.** Other end-of-the-world accounts: The Aztec Sun Stone pictures four previous creations, each terminated by a different natural disaster for which dates are given, as discussed in the text (p. 120). We live in the fifth creation, symbolized by the four-part glyph at the center. (A. Aveni, *Empires of Time*, rev. ed. [Boulder: University Press of Colorado, 2002], fig 7.3)

can ever witness. The modern era has added cosmic collisions to the mix.

Some world religions are less specific about millennial-like turn-overs; others employ different measures of time to clock moments of great change. For example, although its origins are culturally con-nected to Christianity, the religion of Islam anticipates a different end of time as we know it. In the tradition of Mujaddid, a branch of the Shi'ite Muslim faith, every century's end awaits a divinely inspired individual (Al Mahdi) who will usher in the golden age. Some Buddhist sects in China and other parts of southeast Asia look forward to the return of a future Buddha who will come down from heaven at a time when the moral laws and universal truths of the original Buddha have begun to fade away. Such expectations were particularly popular among peasant rebellions seeking a new social order. Perhaps the most famous example of non-Western cyclic turnovers happened when supply planes arrived in the Pacific Islands during World War II. Many natives thought the gods were delivering special goods for them from heaven. They set up symbolic landing strips and warehouses and they conducted elaborate rituals in order to receive these goods properly. "Cargo cults" emerged as part of an old traditional belief in the island religions that a new age was about to be initiated.

Thus, there is nothing unique about Maya cyclic creation. Like so many of the other world age–based calendars, time has a starting point somewhere around a few thousand years BC: 4713 BC in the Julian calendar; 4004 BC in the Western Christianity "long count"; 3114 BC in the Maya Long Count; 2850 BC in the Indian; and 3100 BC in the Chinese calendar. In the short run, human affairs, indeed all aspects of life, seem disjointed. All humans need a deep-time framework to anchor their lives, a foundation that guarantees long-term stability. What better way to fix life's seemingly meandering course than to append the happenings that mark its way by the regularly appointed stations in the durable and dependable long cycles manifest in the heavens? This is where the precession of the equinoxes (the longest of the long celestial cycles), aided by widely held traditional beliefs in astrology and our human obsession with segmenting time by number intervals, enters the picture.

> To wish that the world were other than it actually is can be
> a harmless exercise . . . [building] a new age of peace, justice
> and equality . . . yet it is a common feature of Utopianism that
> those who conjure up such fantasies commonly believe that
> their validity extends beyond the purely personal.[3]

So begins a commendable work on millennialism by historian Nicholas Campion. Although precisely timed end-of-the-world scenarios may preoccupy contemporary popular culture, Americans surely did not invent the myth of time's end and global renewal. The doctrine of world ages goes back at least to the fourth-century BC writings of Plato. Distrustful of humanity, Plato believed that to attain order in the world people needed to be controlled. Nature had a moral mission to provide periodic disasters—fire, flood, earthquake—to cleanse the world of misguided human corruption and give us all a fresh start.

Platonic re-creation was signaled by a return of the planets to their original positions. The Romans, and later Islamic culture, picked up on this idea, each with its own claim regarding which planets needed to align with what astrological sign(s) in order to initiate universal cyclic cataclysms. Christian Armageddon was simply one version of an old pagan idea.

Early Christian Gnostics tapped into the global destruction myth as well. Their history is well worth discussing in the context of 2012 prophecy. Today we are familiar with the term "gnostic" largely through its negative popular form "agnostic," which has come to mean someone who has not decided, or never can decide, the issue of belief in a deity. But as second-century AD orthodox literature implies, the Gnostics constituted a cult of Christian heretics whose beliefs rested on two basic principles. First, the world was created not by a supreme benevolent god but instead by forces intent on imprisoning human souls in physically corrupt bodies. Second—and more important for us in broadening the base of 2012 prophecy—hidden wisdom, knowable only to a select group of adepts, must be employed to achieve salvation or escape from the present world.

True Gnostics sought the same union with God advocated by more mainstream Christians. They felt, however, that it was not sin but ignorance that cut us off from our creator. Gnostic life, therefore, consisted of the search for true knowledge. They believed that revelation could be found among *all* civilized nations and that every faith contained a germ of truth that culminated in Christ. In a very real sense Gnosticism was a form of religious internationalism, which, I think, gives it a lot of influence with today's egalitarian mind set.

Early church fathers viewed Gnostics as dangerous people and they took great pains to discredit them. Simon the Magician, a Gnostic of the first century AD, was baptized Christian after witnessing miracles of healing conducted by the apostles John and Peter. So desirous was Simon to acquire these powers that he even offered money for them. In New Testament Acts, the apostle Peter tells the story:

> But there was a man named Simon who had previously practiced magic in the city and amazed the nation of Samaria, saying that he himself was somebody great. They all gave heed to him, from the least to the greatest, saying, "This man is that power of God which is called Great." And they gave heed to him, because for a long time he had amazed them with his magic.[4]

Yet, God's true miracles amazed Simon. And so, he appealed to Christ's apostles:

> Give me also this power that any one on whom I lay my hands may receive the Holy Spirit.

But Peter rebuked him:

> You have neither part nor lot in this matter for your heart is not right before God. Repent therefore of this wickedness of yours and pray to the Lord that, if possible, the intent of your heart may be forgiven you.[5]

Simon thus became the symbolic founder of Christian heresy and the crime named after him—simony—came to be regarded as one of the most despicable sins against the Christian Church. The lesson, of course, is that there is only one way to make miracles happen: through faith.

Gnostics were magicians of a sort. They reasoned that God, who created everything, also was responsible for the evil in the world. There is a whole world of spirits between us and Him and it is out of *their* sin—not ours—that the world arrived in its present corrupt condition. The only way to seek salvation is through the adept (today we might use the word "psychic"), the one with that latent capacity for true knowledge, or at least someone capable of being trained through rigorous practice. Only (s)he can redeem those among us who can be saved. The rest—the purely material people of the world—are doomed. Only the adept can come to know the magical password needed to ascend the ladder of the demon-filled planetary spheres toward heaven and redemption.

Magic numbers and formulas are paramount in Gnostic thinking, which appeals to the contemporary lover of puzzles. One sect, for example, believed that our connection with the real world derived from the fact that John the Baptist's thirty disciples (one for each year in the life of Christ up to the time he began his ministry) equaled the number of days in the moon's phase cycle and also the number of eons of Gnostic teaching that had elapsed in the world.

To many of us, the magic formulae, the secret words and numbers of the Gnostics, are nonsensical concatenations that the rational mind tends to discredit out of suspicion that they were made up on the spot. If Gnosticism appears to defy common sense, that may only be because our modern outlook has conditioned us not to interpret the Bible literally. For the Gnostic, secret mysteries enveloped in cryptic codes are tied to the notion that knowledge can be acquired directly by revelation and that we all have the power within us, through rigorous self-discipline, to penetrate the confused outer layers of reality and get at the inner meaning of things. Such ideas have great appeal to followers of contemporary popular neo-religious cults. For example, Scientology, or Dianetics, is essentially a democratic form of Gnosticism, which accounts largely for its current popularity. Anyone can access higher knowledge of how to erase traumatic effects through mind control and a well-defined praxis. Once your reactive mind achieves a state of "clear" you can access Scientology's upper limits. Like Raëlians (a cult advocating intelligent design through extraterrestrials), Scientologists employ a lot of scientific terminology and concepts; for example, "engrams," which encode psychosomatic problems. Also many of their rules of practice sound like natural laws. Awed by the power of science, neophytes often are attracted to Gnosticism by the scientific elements they perceive in it—the quantitative element, the numbers and calculations, that may offer a keys to decoding life's complex puzzles.

The magical teachings of early Gnosticism were driven by a yearning for identification with the divine. And they often operated in a climate not unlike the emotional, ecstatic, and reactive environment we find in today's popular Pentecostal movements and the mega-churches of television evangelism. Plotinus, the Neoplatonist philosopher of the third century AD, was witnessed by one of his students to have achieved ecstatic union with God—the "divine in the All"—on at least four occasions. Although its popularity waxed and waned, Gnosticism continued to thrive as a Christian shadow cult, appropriately labeled "occult" by its detractors, through the Dark and Middle Ages. (It is interesting to note the use of the word "dark," associated with evil/negative, here, as opposed to "light," the good/positive.)

One of the focal points of Gnosticism is secret knowledge, and another is concerned with making use of that knowledge to fix dates for the end of the world as we know it. For the literal reader, the book of Revelation in the New Testament is the most widely interpreted Biblical source of the end of time for humanity. All of us will suffer the pain brought upon us by the Antichrist, the "Beast" variously described in Revelation. Only after this lengthy period of persecution, known as the Tribulation, will true believers in Christ rise to kingdom come. What makes Revelation such engaging reading for apocalypse watchers is that it is filled with numbers that can be manipulated to time cataclysmic events—like the celebrated 666, the number of the Beast—and vivid descriptions of visions of the destruction of the world revealed to the apostle John (hence the title of the book). Images related to the apocalypse appear prominently in Christian art of the Middle Ages (for example, see Figure 18). The so-called Alpha and Omega labels used by John to invoke the creation and destruction of the world in Revelation also appear in written form on sarcophagi and murals. There is not space here to chart the full course of Western apocalyptic history; for a thorough account of how the scriptures influenced Western civilization, I recommend Jonathan Kirsch's *A History of the End of the World*.[6]

In a tidy little book titled *Century's End* (originally written in the late 1980s in response to the growing millennial hype over Y2K), historian Hillel Schwartz points out that the Western philosophy of deep time is also characterized by a curious arithmetic based on 1,000-year intervals used to mark key points that lead up to Judgment Day, each portraying the ever-increasing sense of doom that will await the nonbeliever. Perhaps like the Maya, our love of numbers, our calendrical punctuality, and our mentality for exploring large time intervals affect our interpretation of history. As Schwartz wrote just short of the end of the twentieth century, "how could there help being terror . . . at the close of a millennium?"[7]

Why 1,000-year intervals? There is no natural cycle that begets the base-10 system. But there is an obvious simplicity in mapping time in digital units—as the fingertips on the ends of our hands suggest. When humans took to writing to supplement gesturing, it was only natural to carry over this system. When time gets too difficult to manage with a single body, you multiply bodies—tens, hundreds, and so on. (Recall that the Maya likely invented a base-20 system because they tallied using both fingers and toes.) Like the Maya, we build big cycles out of smaller ones. The apostle Peter

---

18. *(Facing page)* Albrecht Dürer's early sixteenth-century painting shows St. John's vision of the end of the world by apocalypse as told to him by an angel *(left)* and recounted in the New Testament's Book of Revelation. Among the fantastic imagery, often interpreted literally over the ages, is the seven-headed dragon *(right)*. (Courtesy of the Wetmore Print Collection, curated by the Art History Department, Connecticut College, New London)

must have been aware of such contrivances when he wrote, "[B]e not ignorant of this one thing, that one day is with the Lord as a thousand years, and a thousand years as one day."[8]

Millennium and apocalypse go together like Christmas and Santa Claus. I think it is particularly important to trace this relationship because I believe it is relevant to the way many Y12-ologists interpret the message from the Maya past. The idea of setting up millennial mileposts along the road to apocalypse and assigning particular significance to the sixth millennium after creation originated in the fourth-century AD writings of early Christian sages. The choice of the number six is based on Genesis, wherein the Lord was said to have taken six days to create the world. If, according to Peter, a day is as a millennium, then the present state of the world cannot be altered until 6,000 years elapse. Wrote one sage, "Let the philosophers, who enumerate thousands of ages from the beginning of the world, know that the six thousandth year is not yet completed, and that when this number is completed the consummation must take place."[9] Then follows the promise of the turning of the great cycle (for believers only): "[A]nd this condition of human affairs [will] be remodeled for the better."[10]

About the middle of the fifth century AD, St. Augustine parsed out this historical chronology of the six successive ages or dispensations:

I   Adam to Noah;

II   Post Flood to Abraham, the father of all nations;

III   Abraham to King David;

IV   David to the Babylonian captivity;

V   Migration out of Babylonia to the birth of Christ;

VI   Christ to the Second Coming, the period in which we live, which points toward eternal rest with God in Age VII.

But this is a *relative* chronology. We cannot know when the end will come without some knowledge of exactly when creation took place.

Medieval chronologists consumed a lot of energy grappling with the issue of absolute chronology; that is, in what numbered year in the BC/AD Christian calendar did the creation happen? The most widely accepted answer to this vexing question flowed from the pen of Archbishop James Ussher (who was introduced in Chapter 4). In his 1650 *Annals of the Old Testament, Deduced from the First Origin of the World,* he tabulated all the begats in the Old Testament and projected them backward from the birth of Christ, which he pegged at 4 BC according to Christian chronology. He placed that event sometime in the reign of King Herod. Ussher's final answer: the world was created by God in 4004 BC. Then for convenience the good archbishop fudged the numbers a tad by rounding things off, thus arriving at the conclusion that the present creation will last from 4000 BC to AD 2000. No wonder Y2K was such a big deal!

If the Second Coming is geared to a millennial counting scheme, perhaps the overturning of sub-cycles, such as centuries, offers prophecies about the ultimate outcome at time's end. Maybe there are signals that can tell us how to prepare. Fin-de-siècle (cycle, or century-ending) prophecies, many of Gnostic origin, fill our history books going all the way back to AD 999. Practically all of them signal decay, followed by destruction, and then regeneration. To counter all this negativity, beginning in the 1290s, Christian prelates organized century-ending celebrations called Jubilees. Spinoffs of fifty-year time markers in the Hebrew calendar, these were intervals of recapitulation, purification, and restoration (not unlike like the Maya *katuns*). Jubilees focused especially on the convergence of meaningful events and of signs in nature and also on the hopes and prophecies of a new age of spiritual reformation—all part of a program dedicated to the notion that transformative happenings were tied to the numbers that marked time's passage.

In Jubilee years penitents marched through central Europe, flagellating themselves, and masses of pilgrims traveled to the Holy Land. The stoked-up fires of spiritualism brought about visions. Comets and other portents appeared in the sky, and former emperors were said to have been resurrected from the grave. In the more somber Jubilee of 1900, which was marked by a huge Sacred Year Exhibition in Vatican City, Pope Leo XIII wrote in his encyclical, "The close of the century really seems in God's mercy to afford us some degree of consolation and hope."[11] This attitude remains with us today in the curious juxtaposition of Armageddon and "joy to the world transformed" that attends alleged megacyclic overturns like Maya 2012.

Are the world ages of humanity fixed in the stars? I intend to explore the resurrection of this theory in American culture in my next chapter. The strong dependence that many diverse Great Year calculations have on movements in the zodiac has played a significant role in creating the so-called monomyth, or the speculation that there may be a grand celestial period recognizable by all humans that underlies all calendars, and that somehow everyone is subconsciously driven to pattern history's great turning points after it.

As we learned in the previous chapter, the precession of the equinoxes is the time it takes a predefined starting point on the zodiac (astronomers have chosen the vernal equinox) to make a complete circuit through all twelve member constellations—a period of 25,770 years. I concluded in Chapter 5 that although any culture attentive to the sky could detect the effects of precession, computing the full cycle is quite another matter.

I think the Christian world age calendar got tied to the precession cycle because of a curious coincidence that can be found in the numbers. The time it takes the vernal equinox to move through one constellation of the zodiac is one-twelfth of 25,770, or approximately 2,150 years—2,147.5 to be exact. In this convenient scenario, one "age" is approximately double the old Augustinian millennial creation era. This leads to the notion of star-fixed world ages, each portending, according to commonly held astrological dictates, millennial or bimillennial portents of joy or gloom. Thus, the Christian era opened when the sun moved into the constellation Pisces (the symbol of Christ in the Roman Catholic Church) at the spring equinox. As Virgil put it, "A new great order of centuries is now being born."[12] The previous age, that of Aries, as Figure 18 illustrates, was marked by Moses's arrival on Mt. Sinai. Before that, when the equinox sun resided in Taurus, people worshiped the golden calf, or Taurus the Bull, symbolized in the headdress of certain Egyptian deities and in the Minoan architecture of the palace at Knossos. The age just ahead of us is quite familiar, at least to my older readers. A popular song from America's 1960s revolution extols the peace and love that will blossom in the forthcoming Age of Aquarius (not slated to begin until the equinox sun enters that constellation, in about AD 2700).

---

To sum up, millennialism, the idea that thousand-year intervals mark big changes in the world, is part of a long Western tradition. Date setting by manipulating numbers is a key part of this tradition, and many of these numbers emanate from Biblical texts concerned with what will transpire at time's end. I think this is what is behind a lot of the Y2K and Y12 hype.

I tried in this chapter to place Y12 in the broader context of ends of time in world cultures, and in the next chapter we pick up the threads of the Gnostic way of thinking and the apocalypse and focus on America's deep and abiding interest in end-of-the-world scenarios, especially Y12. America in particular has a long history of Judeo-Christian apocalyptic forecasting, from the hotbed of utopian ideologies and practices in the nineteenth-century "burned-over-district" of upstate New York to a twentieth century filled with dire cosmic warnings. The same readers who remember the Age of Aquarius will also recall Comet Kohoutek, the purported "comet of the century" in the mid-1970s. And who can forget the more recent anticipated cosmic reclamation project that attended the end-of-the millennium appearance of Comet Hale-Bopp, the return of the "alien mother ship" that ended in mass suicide for thirty-eight members of a modern day cargo cult?

## ONLY IN AMERICA

*Novus ordo seclorum.*[1] Inscribed on the back of U.S. dollar bills below a mysterious-looking eye perched on top of a pyramid, this statement declares the advent of "a new order of the ages." America always was a birthplace for new ideas—new beginnings. As Thomas Paine put it, "the birthday of a new world is at hand."[2] To understand how and why this image of America was cultivated and why the United States has become the spawning ground of so many contemporary apocalyptic theories, including Maya 2012, we need to examine developments in apocalyptic creation history discussed in the previous chapter.

Martin Luther, the famous leader of the Protestant Reformation in Germany, made frequent use of apocalyptic rhetoric in his sermons (he was fond of calling the Pope the Antichrist). His words spawned a millennialist movement that developed sturdy roots in England, from where the Pilgrims and Puritans, who settled in the New England, came. Diggers, Levellers, Ranters, all were early seventeenth-century cults who claimed that the descent of a New Jerusalem from heaven was just around the corner. In anticipation, colonists patterned their cities after the New Jerusalem described in Revelation (21:2, 16, 18):

> Coming down from God out of heaven
> ... the city lies four square and the length is as
> Large as the breadth.[3]

Founded by Puritans, New Haven, Connecticut (my hometown), one of many New World cities conceived as a millenarian paradise, is built around the prescribed grid plan of nine blocks in the form of a perfect square.

The Great Awakening, a religious revival movement developed in the colonies during the 1720s, 1730s, and 1740s, advocated emotional religious experience acquired by mass worship and public weeping. Sermons laced with strong language were designed to terrorize sinners and to root out any false works attributed to God. Some historians have argued that the militancy behind this movement may have attached itself to the brand of patriotism advocated by the American Revolution. But the American version of Biblical time's end came to be viewed largely optimistically, as we see it characterized on our dollar bills. As the influential Boston preacher Cotton Mather wrote, "God surely intended some great thing when he planted these American heavens and earth."[4] Mather is also responsible for a famous statement that would resonate in the heads of countless immigrants for generations to come: "Even though the world would be destroyed by subterraneous combustions and amassments of igneous particles,"[5] after the great firestorm New Jerusalem will descend from the heavens— "a *city*, the street whereof will be *pure* gold."[6]

New England Puritanism developed the generally socialized attitude toward new beginnings that characterizes the prophets of the 2012 phenomenon. In the remote forests of the New World and separated from the Church of England, the Puritans of the Massachusetts Bay Colony codified a test for full membership in the church and for the right to vote in the civil polity as early as the 1640s. The test was based on telling individual conversion experiences. These were public testimonies in which confessors claimed justice in their holy damnation for sins they had committed. Sinners submitted to their eternal damnation and admitted their inability to earn a state of grace by themselves. They publically pronounced themselves ready for ecstatic conversion, fully open to receive whatever peace God might bring them.[7] By the time of the Great Awakening a century later, that experience had become socialized into a general feeling of impending disaster, which contributed to the millennial attitude that would permeate the coming American centuries.

America may have started out as a refuge for Pilgrim and Puritan ideological outcasts from Europe, but the early nineteenth-century American frontier—upstate New York, western Pennsylvania, Indiana, and Ohio—attracted those castaways whose beliefs could scarcely be tolerated in Protestant New England. The more adventurous among them fled the urban blight with its industrial pollution, opting for rural pastures, clean air, and a general clearing of the mind. These restless adventurers, opportunists, and theologically disgruntled zealots made their way west along the newly constructed Erie Canal toward virgin territory, carrying with them their conceptions of a new freedom of thought, accompanied by a revival of spiritualism heavily laced with an apocalyptic vision of the world.

Historians call the area south and east of Lake Ontario, where I live, the "Burned-Over District" because so many oddball Yankee social and religious movements swept over it. Like an overused cornfield, the land reached spiritual burnout by the mid-nineteenth century, when a general westward expansion of the U.S. population set in. Behind this new spiritualism lay the post–Revolutionary War period of economic depression. In the face of persecution, religious extremists in this unsettled new American society were looking for new places and new identities.

Founder of Mormonism Joseph Smith, for example, claimed to have received the Golden Tablets about two-thirds of the way up the canal, near Rochester, New York, in 1827. The forces of persecution drove him and his followers to Ohio, then to Illinois, where attempts to establish a utopian community were met with hostility. Finally, the community settled and took root at the edge of an immense lakebed in the middle of desolate salt flats in the heart of the Rocky Mountains of Utah. Interestingly, Mormonism became the only major active religion to spring up in America.

The Burned-Over District also gave birth to the women's suffrage movement. Early advocates for women's rights Elizabeth Cady Stanton and Susan B. Anthony authored their first proclamation in Seneca Falls, New York, only fifty miles from the wellspring of the Mormon faith.

Lesser known fringe cults sprouted utopian settlements in New Harmony (Indiana), Amana (Iowa), and Oneida (New York). The Owenites, for example, were founded in 1825 by Robert Owen, whose passion to eliminate poverty led him to organize socialist communities, such as the one in New Harmony, consisting of 500 to 3,000 people who lived together in a single building. From jobs to education, these communities were to be entirely self-contained. Protesting what they felt to be the arbitrary rule of church and state, a group of German Pietists known as Inspirationalists had immigrated to Buffalo in 1843 to found their own "Community of True Inspiration." Like the Mormons they were driven farther out onto the frontier, finally settling in Amana. And in Oneida, John Noyes founded his community in 1848. Noyes believed that perfection, or "sinlessness" as he called it, happened directly upon conversion rather than in the afterworld. The Second Coming, he said, had already occurred in Biblical times, and it was up to us to bring about the millennial kingdom through our practices, which included open marriage.

Other sects with strong millennial underpinnings included the Millerites and the Shakers. The Millerite sect of the Adventist Church was founded by William Miller, an influential New England preacher. His incomprehensible additions, subtractions, multiplications, and divisions of Biblical time periods (see Figure 19) con-

**19.** Pastor William Miller's chart depicts calculations that led to his prediction of the Second Coming in 1843. Once the date passed, the Millerites wept in great disappointment . . . which led to several recalculations. (P. G. Damsteegt, *Foundations of the Seventh-day Adventist Message and Mission* [Grand Rapids, MI: Eerdmans, 1977], 310)

verged on an anticipated pre-twentieth-century Second Coming, his so-called Great Awakening, scheduled to happen between the equinoxes of 1843 and 1844. Avid Millerites quit their jobs, sold their worldly possessions, and even confessed to their wrongdoings.

They gathered on rooftops and hilltops garbed in robes, awaiting Jesus to come down from heaven and lift them up. Nothing happened. Dejected and disappointed followers wept for days; some took their own lives. The Great Awakening became the Great Disappointment. Later, one of his followers claimed to find an error in Miller's calculations and reset the date to October 22, 1844, the same day of the year Bishop Ussher had assigned to the creation of the world.

The Shakers, founded by Mother Ann Lee, sought the Second Coming in the form of their leader, whom they considered to be the embodiment of Christ (the son of a bisexual God) in human form. Withdrawing from society altogether, they lived in an isolated community and by their own ethical code, which included full equality for both sexes. They acquired their name from the practice of spirit communication through the rattling of furniture. During Shaker meetings, when spirits of the dead possessed them, in a state of ecstasy (literally, an out-of-body experience) they would shout and howl uncontrollably.

Occult spiritualism saw its heyday in the Burned-Over District as well. The Fox sisters, self-proclaimed clairvoyants, communicated with the dead through spirit rappings. Maggie and Kate's demonstrations of Morse code–like communications with the departed set off a wave of interest in the occult that swept over urban America. Poet William Cullen Bryant, writer James Fenimore Cooper, historian George Bancroft, and encyclopedist Charles Dana were all enthusiastic witnesses. Horace Greeley was so taken with the teenage country girls that he volunteered to pay for their schooling. (The girls later confessed to making the knocking noises by clicking their knee and ankle joints.) Table tipping, séances, spirit writing, and self-levitation followed. The American fad of séances even was exported to Europe (they were particularly popular in France).

As I have noted elsewhere, revelations acquired in this pre–Civil War American religious fervor often tended to be framed in terminology that reflected the period's scientific breakthroughs.[8] The early nineteenth century had been an exciting period of discovery of new forms of energetic disturbances. In 1831, for example, physicist Michael Faraday demonstrated that passing a magnet through a coil of wire generated electricity. Franz Anton Mesmer co-opted Faraday's ideas, proposing the theory of animal magnetism—that is, that all living beings possessed a magnetic fluid. In his expert hands (so he claimed) this magnetic fluid's healing power could be manipulated to cure diseases, blindness, sexual inhibition—even boredom. The invention of electric motors and generators soon followed, along with Morse's telegraph, which became the perfect metaphor for communicating with dead spirits by rhythmic tapping.

The nineteenth-century pop-craze for supernatural prognosticating became so rampant in America that the U.S. Congress entertained a bill in 1854 that proposed to organize a national committee to investigate "certain physical and mental phenomena of questionable origin and mysterious impact that have of late occurred in this country."[9] Nearly 100 years later, in 1966, that august body would take similar action to investigate alleged sightings of unidentified flying objects. By this time, it seems that Biblical apocalyptic ideas had been transformed to meet the needs of a more secular America.

Employing a little-known early sixteenth-century secret document, a fringe group based in England known as the Hermetic Brotherhood of Luxor calculated that the world as we know it would cease to exist in the year 1881. This calculation was based on the theory that each of seven angels (representing the five visible planets and the sun and the moon) would rule the universe twelve times during one "Great Solar Period," or the time it took the equinox sun to pass all the way around the zodiac—the period of the precession of the equinoxes now familiar to us all. At the end of that cycle, the world would end. Other adepts who followed the cult mathematically calculated ends-of-time in 1879, 1880, and 1882. "As to the series of events that will then take place, that would take a prophecy," wrote one sage.[10]

The late nineteenth-century flourishing of science emboldened more than a few rational skeptics. One, for example, thought the 1881 end-of-the-world prediction was an attempt to persuade Anglo believers that they were the true descendants of the Lost Tribes of Israel. We are all "victims of 'suggestion,'" wrote the critical French historian René Guénon.[11] Assume for a moment that the Egyptians really did build mathematically based prophecies into their pyramids. Could it be, Guénon asks, that they possessed a knowledge of the relationship between the history of the world in general and the cycles in which that history is framed? But, added Guénon, would it not have been in their interest to adapt that history to their own culture? And would they not leave evidence in the form of dates and events in dynastic history that fit those cycles? Why then is all the information we extract from pyramid data framed in Judaism and Christianity? Is the Great Pyramid a Judeo-Christian monument? And why has no data of this kind been recorded since antiquity? It may be comforting to think our place in the world has deep roots, but consider the burden the precisely timed prophecy of humanity entering a new era and the advent of a great spiritual renewal places on our ancient ancestors.

Self-interest, Guénon further argues, is another motive for predicting the timing of a cataclysmic event. Announcing that a revolution will take place at a particular time assists those interested in its breaking out at that very time. He adds: "Certain people want to create a state of mind favourable to the realization of 'something' that is part of their plans; this something can no doubt be modified by the action of contrary influences, but they hope that their methods will serve to bring it about a little sooner or later."[12] Guénon's arguments still make sense today.

What was America like when the 1881 and other contemporary end-of-the-world theories were postulated? The Gilded Age had begun in the 1870s. Bustles and corsets confined the feminine body, but the body politic proliferated, with voter turnouts in local elections approaching 90 percent. Not until 1968 and 2008 would America witness a comparable interest in the elective process. Voters had plenty to be concerned about. The Crédit Mobilier scandal of 1872 revealed graft and corruption in the sale of bonds for the Union Pacific Railroad to political cronies. Unemployment in 1877 stood at 15 percent. Farmers who felt victimized by railroads, merchants, and banks formed their own secret society, the Grange. Dissatisfied with both Republicans and Democrats, populist politicians created their own fringe parties—the Greenbacks (dedicated to the expansion of paper money), the Prohibitionists (who waged the earliest battle to ban alcohol), and the Equal Rights Party (advocating women's suffrage). The end of the decade saw a huge financial recession. Still, the world did not end for all of humanity in 1881, although it did for President Garfield. He was assassinated.

Followers of the 1881 and other predictions about the end of the world deployed the oft-quoted words of Old Testament prophet Daniel on the impending cleansing of the world: "Seventy weeks are determined upon thy people, and upon thy holy city . . . then shall the sanctuary be cleansed."[13] Because days are often interpreted as years by many doomsday forecasters, Daniel's statement is sometimes construed to refer to the "one day of precession" discussed in Chapter 5.

The impending apocalypse is portended by visions of horrific beasts that he describes: "[T]he visions of my head alarmed me. I approached one of those who stood there and asked him [the angel with whom he is conversing] the truth concerning all this. So he made known to me the interpretation of the things."[14] But few literal interpreters of the Bible bother to pay attention to what else Daniel has to say: the angel replies that the beasts who will arise out of the earth are only symbolic. "These four great beasts [who arise out of the earth in Daniel's vision] are four kings . . . and . . . as for the fourth beast, there shall be a fourth kingdom on earth . . . an evolutionary kingdom, and all the dominions shall serve and obey them."[15] At least Daniel seems to have had a firm grasp of literary symbolism.

In the New Testament version of the apocalypse (literally, the "unveiling," from the Greek work *apokalypsis*) that appears in the more well-known Book of Revelation, it is the apostle John who tells of the encounter with an angel who reveals the secrets of the "last things." In content it is very much a replay of what appears in Daniel and, according to most Biblical scholars, it never was intended to be comprehended literally. The New Testament version tells of the Rapture, when Christ will miraculously come down from heaven and pluck his followers off the face of the earth, just in time to become heavenly spectators with a ringside seat from which to view the destruction below. Then follows the Tribulation, a time of suffering under the reign of the Antichrist. A great battle between the forces of good and evil will then occur at a place known as Armageddon (possibly a corruption of Megiddo, a strategic outpost in the Holy Land where many real battles were fought). Finally, New Jerusalem, the kingdom of God, will be established as a paradise on earth for eternity. As in the case of Daniel, a vast majority of scholars acquainted with the apocalyptic concept argue persuasively that Revelation is more concerned with symbolic lessons than with secret codes that pertain directly to the fate of humanity.[16]

It may be only a coincidence, but many of the late nineteenth-century dire predictions, including that of 1881, were issued on the heels of the 1859 publication of Darwin's *On the Origin of Species*. Calls for the cleansing of the world and the dawn of a new consciousness were common utterances from the mouths of other end-of-the-world prophets, such as the celebrated seer and mystic Helena Blavatsky. Her nineteenth-century American version of cosmic consciousness conceived of a growth process in which life passed upward through progressive levels of existence. First came an invisible race made of fire and mist that lived in the polar region; then came a race of red people who inhabited northern Asia (and are responsible for sexual intercourse). There followed, respectively, the people of the lost continents of Lemuria (in the Pacific) and Atlantis (in the Atlantic). We are the fifth race, but there will be others and they will grow increasingly more spiritual as humanity steps through time to the beat of an infinite, eternal deity who pulls us ever forward up a ladder of progress. Blavatsky's world age model can be characterized as a theory of "assisted evolution" not unlike some of the 2012 ideas (e.g., Calleman's) I reviewed in Chapter 2. In effect, Blavatsky replaces the out-of-step God of the Old Testament with a "hip" deity who is in tune with creation and evolution by natural means—a scientific god. This was a nifty way of resolving tension between religious faith and Darwinian science—at least to the superficial reader of Darwin.

I see similarities between the contemporary scientific creationist's reaction to Darwinian evolution and the apocalyptic prophet's feelings about 2012. Both center around the way science has changed what time means. Fueled by a distaste for the notion of descent from an ape, the former are basically concerned about the origin of humanity, whereas the latter, unable to cope with the extraordinary duration of time the tenets of science offer them, direct their attention toward predicting humanity's fate within a more finite time period. There are other parallels. Historian Ronald Numbers notes that scientific creationists, who have enjoyed enhanced popularity since 1980, have traditionally been regarded as purely American.[17] Like Y12-ologists they appeal to scientific terminology even as they reject the findings of establishment science.

In addition to secret Bible codes, those who have sought clues to our future in antiquity have looked to the pyramids. These structures also exhibit a history of offering a novel spin on the Gnostic notion that secret knowledge is encoded in records that go back to antiquity. Piazzi Smyth, Scotland's astronomer royal of the late eighteenth century, spent a good part of his time in Egypt carefully tabulating measurements (in "pyramid inches") of galleries, chambers, and passageways in the Great Pyramid. One of his examples of how to use a correspondence between numbers to discover hidden prophecies related to the mythology of renewal points directly to the 1881 date I referred to earlier:

> If you let fall a plumb-line from the entrance of the way of
> escape at the S.E. top of the Grand Gallery, it will intersect
> the top of the great step. So that then, instead of continuing

an imaginary line of floor distance measurement through the step, measure up the step, and along the top to the spot where the plumb-line would intersect, and I fancy the measurement would be 1813 + 36 + 31.2 = 1881.2 [*sic*].[18]

The Egyptians were not the only pyramid builders consulted on the matter of last things. In the late 1970s the "pyramid power" craze—pyramidology—resurfaced, this time with the Maya in the starring role. "Was the emblem of the pyramid adopted for the seal of the United States because it represented the structure of the world. . . . Certainly [its choice] is validated by the pyramids of Mesoamerica," wrote mystic Frank Waters.[19] American devotees responded by climbing into homemade versions of pyramids, personal polygons you could assemble from a kit. Once you got inside you could sit in a lotus position at the power point in the center. What happened next?

> I felt my entire body vibrating . . . a feeling of quietude and
> relaxation . . . clairaudient, sensing the sounds of the woods . . .
> [then] strong vibrations . . . a bright white light . . . an effortless
> flow of energy.[20]

Many aficionados of pyramid power believed the ancients were able to create crystals and pyramids because they possessed high-tech science well beyond the levels of our own. Likewise, today's 2012 prophets have focused on the physics of magnetic fields and black holes at the center of the Galaxy. The story they tell centers on the notion that our skilled predecessors, like us, possessed the technological capability of doing themselves in. Aware of this dilemma, they decided to conceal their powerful knowledge in geometrical forms, a record we can now use to recover the potential to change ourselves and the world. We just need to wise up and tap into their ideas.

Peter Tompkins's flighty *Mysteries of the Mexican Pyramids*, which reached press just after Waters's influential book, extolled Mesoamerica's pyramids. It also made subtle connections between

Maya mythology, outer space, and Christian apocalyptic tradition. Concerning the plumed serpent, for example, he writes, "Moving and undulant, the serpent in Mesoamerica symbolized life, power, planets, suns, solar system, galaxies, ultragalaxies, and infinite cosmic space."[21] For Tompkins, the pyramids on which their effigies were carved symbolized man's ascension from the dark subterranean domain of Tezcatlipoca upward toward heaven and the light and wisdom of Quetzalcoatl (the feathered serpent deity). Scenes from the codices depict Quetzalcoatl descending from the firmament on a cord (Figure 20). Thus, the descending serpent and the descending God of Christianity emerge as distinctly parallel, at least on the surface.

In 1975, just about the time the new wave of fascination with the Maya began to take hold, I was finishing some survey work on the possible astronomical alignments in the Caracol at Chichen Itza, a round structure with narrow windows in its turret. I recall repeatedly running into Luis Arochi, a Mexico City journalist. As we elbowed each other out of the way, shooting pictures of the equinox sun through one of the Caracol's windows, Arochi told me that he had come to Chichen Itza to look into the then practically unknown serpent hierophany. A year later his book on the phenomenon initiated the popular spring equinox pilgrimages to the most famous of all Maya ruins to watch the serpent descend from heaven (see Chapter 2 for my eyewitness account).[22]

It is no accident that both Tompkins's and Waters's books saw the light of day at the same time Maya epigraphers were making an accelerated assault on cracking the Maya code.[23] Like Joseph Smith's golden tablets or the "three secrets" received from on high in Portugal's miracle of Fatima,[24] a possible source of secret knowledge concerning universal truths from the distant past was just beginning to come to light. Tompkins's book in particular, with its cryptic passages from the works of nineteenth-century adepts and quaint antiquated photos and diagrams, garnered a lot of attention. What better way to sell books than to tickle a reader's fancy with secret knowledge.

**20.** As in the Christian Second Coming, a god descending from heaven appears as a common theme in Mesoamerican art. In this scene from the Vienna Codex, Quetzalcoatl (Kukulcan) comes down from the sky on a cotton rope. (Vienna Codex, page 48, Akademische Druck-u. Verlag, Graz)

---

Also contributing to the upswing of interest in millennial thinking in the 1970s was the 1970 publication of an influential prophetic novel by apocalyptic writer Hal Lindsey, *The Late Great Planet Earth*, which sold 20 million copies. Lindsey's plain-talking narrative repackaged Revelation from start to finish in the Cold War context of the times, with America enveloped in the terror of world annihilation by nuclear weaponry. Lindsey's Antichrist was a Soviet political leader and Armageddon was the final attack on the cities of the world by ballistic missiles. Lindsey's chronology targeted 1981 for the appearance of the Antichrist, followed by seven years of Tribulation, and culminating with the Second Coming in 1988. Needless to say he was wrong, but like most end-of-the-world time fixers, he gave himself a second chance in two later works, the last one published in 1994. In it, Islamic fundamentalists replaced the godless Soviets in the role of the Antichrist. Readers again were given the apocalypse with which they could identify.

Although the secular American seed of the apocalyptic movement was planted back in Puritan times and has undergone several resurrections, today it has become more marked. It received a boost in the 1970s, thanks to President Jimmy Carter's revelation of his born-again faith, spurring a huge surge in evangelicalism in America. Indeed, most presidents since then have declared themselves born-again Christians. And fears about looming apocalypse were further fueled by President Ronald Reagan, who once said, when commenting on the status of world control of nuclear weapons, "[T]he day of Armageddon isn't far off. Everything's falling into place. It can't be long now."[25] The number of Americans who describe themselves as born-again Christians has remained fairly steady since the 1970s. A late 1976 Gallup Poll put that percentage

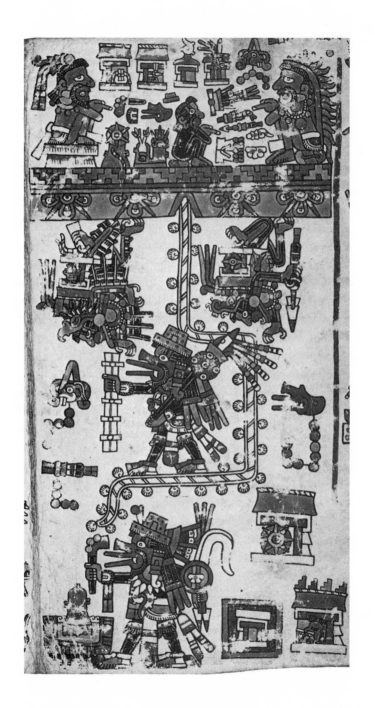

at 34 percent, and later surveys have placed it at 25 percent in 1977, 26 percent in 2004, and 28.6 percent in 2007. Given the number of professed evangelicals, it is not surprising that the sort of spiritual orientation that conditions such beliefs has continued across the millennial divide. Interestingly, the premillennial increased interest in end-of-the world predictions was accompanied by an increasing number of UFO sightings (especially in the 1970s), as well as a fascination with Eastern religions, parapsychology, and other exotic phenomena.

The American view of apocalypse, especially the idea of Rapture, remains very much a part of Hollywood's stake in the 2012 world view. To cite a few examples, the final scene of the cult film *The Rapture* depicts the heroine, a born-again Christian who had committed a terrible crime, and her agnostic lover being lifted into heaven just as they are about to be overtaken by the Four Horsemen of the Apocalypse. *The Omen* series, which dates back to 1978 (at this writing three sequels have followed), focuses on the dilemma of a father who unknowingly adopts the Antichrist child. To save the world from Armageddon he must murder the devil using the seven daggers from the site in the Holy Land where the Biblical Battle of Armageddon took place. The 2005 version pegs the end of days at 6/6/06, that famous diabolical number from Revelation. *A Thief in the Night* (1972) is another cult classic that became the archetype for many Rapture films to follow. In this film, a secular nonbeliever awakens one morning to learn that her Christian husband and millions of other believers have vanished from the earth. The Rapture is already underway and, in a plot line reminiscent of *The Invasion of the Body Snatchers*, those left behind are chased down and marked with the stamp of Satan. Suddenly, she awakens again only to find that it was all a dream. Or was it? Her husband is missing.

Books on Rapture themes remain popular. These include the clever series that began with *88 Reasons Why the Rapture Is in 1988* by Edgar Whisenant, which was followed by *89 Reasons Why the Rapture Is in 1989*, and then similarly titled works in 1993 and

1994. (Make up your mind!) Televangelist Jack Van Impe, who had predicted widespread catastrophe in Y2K, has recently shifted his predictions of the Second Coming to 2012. Fundamentalist preacher Chuck Missler's 1998 video *Alien Encounters* claims to uncover Biblical evidence that leads to identifying the real meaning behind UFO sightings—the Rapture. And Jerry Falwell, among the most celebrated evangelical preachers, laid blame for the impending end of the world on the Middle East crisis. In 1999 he predicted that the Antichrist would arrive within a decade; and "[o]f course he'll be Jewish," said Falwell.[26] He made no mention of the Maya.

Popular interest in premillennial times was accompanied in the scholarly sphere by the appearance of the highly provocative and influential book (especially in the United States—it was published in Boston) *Hamlet's Mill: An Essay Investigating the Origins of Human Knowledge and Its Transmission through Myth* (1969), by science historians Giorgio de Santillana and Hertha von Dechend. "What if we could prove that all great myths have one common origin in a cosmology not terrestrial but celestial?" its authors boldly queried.[27] What if gods and mythic places are mere stand-ins for celestial phenomena? And what if complex astronomical data are the language for *all* world myth? The idea that myth is a preliterate form of science underlies much of what New Age Gnostics seem to be saying about Y12, and I am convinced that ideas such as those proposed in *Hamlet's Mill* lie at the core of much of their prophesying.

Precession of the equinoxes is central to de Santillana and von Dechend's theory of the monomyth, which jibes pretty well with the religious internationalism that envelops all New Age Gnostic thinking. They argue that the passage of the vernal equinox through the constellations of the zodiac (recall Figures 14 and 16a) triggered dramatic changes in the development of world civilizations. Long before Maya 2012ers took hold of this idea, a few other American writers, some admittedly heavily influenced by *Hamlet's Mill*, had applied this idea to a variety of cultures. Some of their theories are worth recounting.

In his widely reviewed book *The Once and Future Star* (1977), writer George Michanowsky claims that writing on a Sumerian clay tablet (the reader never quite gets to learn the details of his translation) proves that the ancient Mesopotamians had once worshiped the Vela X supernova as their ancestors' south polar star. They were able to back-calculate its polar position because they knew about the spectacle. Michanowsky attributes significant parts of Sumerian mythology to this colossal phenomenon that lit up the sky for several weeks sometime between 9000 and 4000 BC. According to him, the "celestial prodigy was eventually remembered as the visit of a supernatural being in semihuman form."[28] "The impact of the Vela starburst," Michanowsky continues, "triggered a development in their culture that, in a short time span made them something entirely different from what they had been before. For better or worse, humanity had thus quite suddenly bitten into the fruit from the tree of knowledge."[29] This impact was far-reaching, Michanowsky writes: "My research indicates that this heavenly event became the source of the creation myths, the cosmological concepts, and the cultural traditions of much of our civilization."[30]

Robert Bauval, an engineer with more than a passing interest in Egyptology (we met him earlier in Chapter 3) also invokes world ages tied to precession. In 1995 he coauthored the popular *Orion Mystery*, published by Crown, a trade-text publisher in New York, in which he claimed that the layout of the pyramids followed a preordained celestial template, with the three largest monuments lining up with the stars in the belt of Orion.[31] Bauval's latest theory advocates that around 2800 BC, thanks to their newly acquired knowledge of precession, Egyptian sky priests used a celestially based religion as part of a plan to overturn the existing order. In his most recent work, *The Egypt Code*, he advertises:

> My new theory does not stop here, for I will also demonstrate in *The Egypt Code* that the slow cyclical changes witnessed in the sky landscape, caused by precession and by the peculiarity of the Egyptian civil calendar over the 3,000 years of the

pharaonic civilisation, are reflected in the changes witnessed on the ground all along the 1,000-kilometre-long Nile Valley in the evolution of temples throughout the same 3,000 years. In other words *The Egypt Code* proposes, no less, to prove that there existed a sort of "cosmic Egypt" ghosted in the geography of the Nile Valley, stretching from north to south, that was once literally regulated and administered by astronomer-priests headed by a sun-king, that lasted for over three millennia, and that can still be discerned in the layout of pyramids and temples that remain today.[32]

I was not surprised to find in Bauval's scarcely comprehensible writings a stance widely shared by many Y12 prophets. He complains that his earlier book had been "met with a barrage of academic indifference." Yet he remains undaunted because he has generated "massive interest and support among the general public and the international media."[33]

Another example of an Anglo-based book on social change triggered by precession concerns Mithraism.[34] Developed in ancient Persia, the Mithraic cult became popular in Augustan Rome, particularly among the military class. It rivaled and almost displaced nascent Christianity. Mithraic iconography is notable for its stark and direct visual symbolism, and in particular for the order accorded the planetary positions. For example, the Mithraeum, or underground temple of worship at Ostia Antica near Rome, also known as Sette Sfere (Seven Spheres), is thought to be a virtual map of the cosmos. The zodiacal signs are laid out in counterclockwise order around the tops of a pair of cardinally aligned benches. There worshipers confronted the god Mithras, whose effigy is surrounded by solar and lunar torchbearers, along with an odd assortment of animals—raven, snake, lion, and scorpion among them—in a bull-slaying scene called the "tauroctony." Planetary signs also gird the sides of the benches. The arrangement makes no sense, unless, as classical historian R. L. Gordon argues, you look at it as the order of planets the way they were perceived in the sky on the night of creation.[35] In fact, the order is the same as that prescribed in astral lore and found on reliefs, for example, at Sidon (today in Lebanon). The inside of the Mithraeum depicts the horoscope of the world, based on what occurred on the night of its birth, when Mithras killed the bull. Mithras was the lord of time and a sun god in ancient Persia, so these roles help explain why he is so often draped with zodiacal figures. But what would a god of time accomplish by killing a bull?

In philosopher David Ulansey's decoding of the tauroctony scene, the animals represent the actual ordering of constellations based on the sun's position in the constellation of Taurus at the spring equinox. The effect we attribute to a mechanical precession, our predecessors would have ascribed to some sort of cosmic deity empowered to alter the star-fixed ages. According to Ulansey, the slaying of the bull by Mithras depicts the central mystery of the cult, the power of their god alone to move the entire universe, for by slaying Taurus the Bull, Mithras allowed the sun at the equinox to move into the next house of the zodiac. In effect, Mithraists sought to use their knowledge of the precession of the equinoxes to devise a religious scheme that would overcome the forces of fatalism.

A final example of world-renewal hypotheses published just before the millennium and specifically based on precession applied to cultures other than our own (and that of the Maya) is William Sullivan's *The Secret of the Incas: Myth, Astronomy, and the War against Time*.[36] In this work the cosmic determinism of *Hamlet's Mill* meets the New World culture of the Andes. Sullivan contends that the world ages of Inca myth that I discussed earlier (see Chapter 6) are all related to transformations in the celestial sphere. He echoes Bauval's familiar scenario that astronomical revolutions—astronomical breakthroughs and new visions of cosmic reality—led to great social revolutions, the most recent being the birth of the Inca empire. Thus, in the most current epoch, the Inca god Viracocha and Manco Capac (first king of the Inca) enter into conjunction in AD 650 as the celestial personages of Saturn and Jupiter. This also happened at other significant times in real Andean history, claims Sullivan. Curiously, the Milky Way also enters the picture. Its shift relative to the heliacal rising of the sun (this time at the June solstice) is a part of the celestial myth as well. Sullivan's extraordinary claim of being able to pin transformative cosmic dates to within a fifty-year time frame is reminiscent of Jenkins's claim regarding the celestial basis of the Maya calendar. (Sullivan also informs us that the Maya and the Inca were in close contact for long periods of time.)

The Maya were not excluded from the Aquarianism that thrived in the revolutionary premillennial age. Frank Waters noted that the precession cycle was more or less evenly divisible by the Long Count.[37] He even took the trouble to find an astrologer to calculate the planetary positions attending the closing date of December 24, 2011, which, he proffered, exhibited many harmonious aspects. But Waters was off by a year. Evidently, he forgot that when you count years in the BC/AD system there is no zero year (thus AD 2 = +2, AD 1 = +1, 1 BC = 0, 2 BC = −1, and so on).[38]

I am convinced that a well-worn, thoroughly marked copy of *Hamlet's Mill* lay on the shelves of all of the aforementioned Anglo authors who promoted precessionally based world-age renewal. Indeed, the opinion of an *Atlantic Monthly* reviewer quoted on the book jacket of *Hamlet's Mill* has turned out to be truly prescient: "It is likely to remain . . . a lion in everybody's path for years."[39]

Thus far in this chapter I have gone to some length to demonstrate that the early Anglo-American version of global renewal through the apocalypse, descended from early Christian Gnosticism, has been around on the American scene in one form or another ever since the first European immigrants arrived, and that it is alive and thriving as 2012 approaches. Little has changed. American political rhetoric has never deviated from the theme of a timed world renewal propounded by Thomas Paine and written on the dollar bill—from Ronald Reagan's "Morning in America," to Bill Clinton's "Springtime in America," to Barack Obama's "Our Time for Change

Has Come." This practically unique American view helps account for the fact that one does not come across many French, Italian, or Chinese publications on this topic.

I have also tried to show that the American fascination with end-of-the-world utopian visions has waxed and waned. As 2012 approaches we seem to be experiencing another upturn. Why now? The same polls that document the high, relatively constant evangelical population also reveal that our culture is beset with fear. In addition to the ever-looming threat of a nuclear holocaust, which has been with us since Hiroshima, the twenty-first-century American angst has been compounded by the 9/11 incident and the threat of terrorism by religious extremists, the arms race, public awareness of accelerated climate change and its related ozone holes and rainforest depletion, and post-millennium natural disasters of extraordinary magnitude, such as Hurricane Katrina, the typhoon in Myanmar, the Indonesian tsunami, the earthquake in China, and so on. And add to all of that bad news a cataclysmic world economic collapse. When we lose our faith in established institutions, as in the 1870s and at other times in history, many of us search for answers elsewhere in the face of calamity.

Historian Hillel Schwartz lists a number of common denominators that have characterized beliefs about the end of the world. These include the following:

- A joyful insistence on the continuity of the generations despite turmoil and threats of impending disaster and chaos looming over our heads.

- The notion that a critical, calculable moment is upon us and that this moment will open the door to global renewal.

- Secret knowledge about the world's end being encoded in calendrical numerology and astronomical conjunctions or convergences.

- In some cases, an appeal to perceived higher forms of knowledge, such as the fourth dimension, out-of-body experiences, imagined aliens, or visions of a future world.

- A polar shift from decay to rejuvenation, from death to purification (here the apocalyptic view of Judeo-Christian Western history comes to mind), stemming from the overturning of great cycles of time.

- Finally, "strange stirrings of delight in the absolute fact that everything must be (on the verge of) falling apart for everything (at last) to come together."[40] We just happen to be here at the right time in history to witness it all.

I would add to Schwartz's list of characteristics of millenarian attitudes the idea that secret knowledge about time's end is buried in written records that emanate *from a distance*. This distance can be spatial, as with distant or lost cultures or even alien civilizations, or it can be temporal, as in the predictions of Nostradamus or the mysteries encoded in the Bible or in ancient Egyptian or Maya script. Conveniently, the ancient Maya's culture is separated from us in both space and time.

Today's 2012 New Age literature, as we have seen, advocates secret wisdom from lost civilizations. Some promotes life-transforming journeys of discovery to faraway places (the so-called sacred travel discussed in Chapter 2). Once we sought a fresh start in communities along the Erie Canal and later on the open frontier of the American West; however, today we seek a higher "galactic" truth. Who can forget the indescribably quizzical expression on the face of Heaven's Gate cult leader Marshall Applewhite as he looked the end of the world in the eye on the April 7, 1997, cover of *Time* magazine? Holding firmly to the belief that Comet Hale-Bopp was about to descend from heaven to transport their souls to some faraway afterlife, he and thirty-eight cult followers had committed suicide in March of that year.

In his introduction to Jose Argüelles's *The Mayan Factor*, Brian Swimme of the Institute in Culture and Creation Spirituality speaks of the "galactic beam" through which our planet is now passing, the "galactic code of the seasons," and our personal interaction with the "galactic mind." Chapters of Argüelles's book bear titles such as "The Galactic Master" and "History and the Solar System: The Galactic View." "The Forgotten Galactic Paradigm" is the title of one of the chapters of Jenkins's *Maya Cosmogenesis 2012*. It evokes a quintessential twenty-first-century version of the star-fixed ages of humanity. Both chapter and book end with a Blavatskian appeal to participate "in the galactic processes of Maya cosmogenesis, recognizing our place in the great chain of creation." Only when we commit ourselves to the "galactic processes" will we ennoble our souls and elevate our spirits "to a plan infused with unity and relationship."[41] Asks prophet Jenkins: "Are we receiving archaic messages through the 2012 time doorway, knowledge to help us evolve? What is it about the galactic center that might make this possible?"[42] For Jenkins the end of the Maya cycle will open a doorway of opportunity to a "conscious relationship with each other and a creative participation with the Earth process that gives birth to our higher selves"[43]—nothing new here.

Galactically derived higher wisdom has even permeated the rational domain of science:

> I fully expect an alien civilization to bequeath us vast
> libraries of useful information, to do with as we wish.
> This "Encyclopedia Galactica" will create the potential for
> improvements in our lives that we cannot predict. During the
> Renaissance, rediscovered ancient texts and new knowledge
> flooded medieval Europe with the light of thought, wonder,
> creativity, experimentation, and exploration of the natural
> world. Another, even more stirring Renaissance will be fueled
> by the wealth of alien scientific, technical, and sociological
> information that awaits us.[44]

These are not the words of a Gnostic New Age prophet. They were written by Frank Drake, the reputable astronomer who founded SETI, the Search for Extra Terrestrial Intelligence, through signals acquired by radio telescopes. Drake's words respond to the question, What will happen on that grand occasion when we finally make contact with beings from another planet? Because we assume, once aliens do communicate with us, that their knowledge must be of a higher form, Drake argues that we will learn from them how to harness and control vast energy resources and how to coexist peacefully. They will teach us the secret of immortality. And we will have a better life. Should we expect anything less from the distant Maya than we might anticipate from faraway extraterrestrials?

Substitute the moment of 13.0.0.0.0 for the time of contact with the aliens in Drake's forecast and you get pretty much the same message concerning our initiation into the galactic club from Argüelles:

> Amidst festive preparation and awesome galactic-solar signs psychically received, the human race, in harmony with the animal and other kingdoms and taking its rightful place in the great electromagnetic sea, will unify as a single circuit. Solar and galactic sound transmissions will inundate the planetary field. At last, Earth will be ready for the emergence into interplanetary civilization.[45]

I asked my friend, psychologist Belisa Vranich, to share her thoughts about why she thinks the word "galactic" resonates with so many enthusiasts of the contemporary end-of-time philosophy of new beginnings and where she thinks the human ego fits into the Galaxy. "To start to understand the galaxy," she wrote me, "you have to tolerate the idea of things that are psychologically incomprehensible and even threatening."[46] She added that to be a part of something limitless is to be threatened, so the idea of anything so vast terrifies us. How can our lives, our chemistry, our selves matter in such vast quarters? As we all grow and mature we go from thinking of ourselves as the center of our universe to realizing that we are not. Within that context, she added, learning that the universe is so large becomes even more jarring. And when we add the possibility of a Big Bang, which we do not have the power to halt, we begin to grasp at straws in our attempts to discover a universe with meaning. Maybe we fancy the galactic concept in the vain hope that in that vast infinite grandeur there might be something for us—some hidden meaning.

Vranich cited a number of studies that show Americans are becoming more narcissistic. We immerse ourselves in activities such as perusing Facebook and watching reality TV shows and Internet Web sites detailing real-time lives of people dieting, dying of cancer, or "finding themselves." We fill our time by employing various cyber-outlets of self-expression, such as text messaging, twittering and tweeting every thought that comes into our heads. Ours is a dummied-down culture with little interest in reading (much less *critically* reading) about anything that might challenge us to think.[47]

I think historian René Guénon's critique about the 1881 end-of-the-world prophecies I mentioned earlier in this chapter holds up today. Cultures that view themselves as waning are usually those most likely to tap into doomsday scenarios, especially when the world's end is followed by universal renewal. What will happen to us? seems to be the sort of question posed by a *reflexive* culture, one seemingly bereft of free will and sharing in the belief that people cannot play a role in affecting the future (the ubiquitous bumper sticker "Shit happens!" comes to mind).

Ours is a techno-immersed, material-oriented society. As a culture we feel uncomfortable about where rational, empirical science is taking us. In a sense, the explanations of the calculated endpoint of Maya time offered by Jenkins, Argüelles, and many other Y12-ologists, with their pseudoscientific cosmic underpinnings, masquerade as science as much as those of the nineteenth-century table rappers and clairvoyants who looked to exotic forms of energy as keys to open the mysterious door to the hereafter. Dissatisfied with the findings of establishment scholars who labor to peer into the Maya ethos, they highjack science to serve their own purposes.

Add America's increasing narcissism to the mix of maverick faith and loss of meaning in the face of a seemingly indifferent, vast universe and you have the perfect recipe for American culture's dizzying rapture with the Maya end of time as 2012 approaches. We desperately seek a better way to live. Conditioned by the Puritanically derived imagination that compels us to view our culture as bankrupt and in dire need of renewal and rapturous uplift, we emerge as a secular people disconnected from our religious roots, yet still fueled by a kind of spiritual quest that grew out of them. Unable to find spiritual answers to life's big questions within ourselves, despite intense searching, we turn outward to imagined entities that lie far off in space or time—entities that just might be in possession of superior knowledge by virtue of having achieved spectacular accomplishments that they may have chosen to encode into the material record for future adepts to uncover.

If patterned great cycles like the Maya Long Count hold the key to anticipating the future, then we must know precisely when the cycle will restart so that we can prepare for the outcome. We need to do the right thing, be in the right place—connect. And so we undertake pilgrimages to ancient Maya places at critical points in their calendar, hoping that their mysterious secrets might have some meaning for us. Yesterday's Lost Tribes of Israel are today's temporally remote Maya. If we cannot talk to ourselves, maybe we can talk to the Other—even if they never spoke our tongue. Evolution needs an assist; our world needs saving. The Maya offer us a transcendent fix, an alignment, an epiphany. And so we romance them.

# EPILOGUE: ANTICIPATION

Preparing this manuscript during a period of world history that some pundits have described as the greatest economic collapse in the lifetime of most Americans has elicited a number of comments from family, friends, and colleagues. Some say, "Too bad the Maya got it wrong by four years" or "Maybe they made an error?" (Oddly enough, Bishop Ussher's calculation of the start of creation necessitated a four-year correction.) Others ask, "What if the great economic downturn is just the beginning of the downslide of a roller-coaster ride that will lead to doom—or Rapture—in 2012?" I cannot speculate except to predict that the dismal outlook forecast by some economists will only exacerbate the feeling of catastrophism that has accompanied the passage of civilization across the millennial divide and toward Y12.

Also at this writing, three years have passed since my e-friend Dylan sent me that alarming missive. Then a frightened young man, he told me of the articles he had read online about great cosmic shifts and the end of the world in 2012. He was particularly concerned about the earth's magnetic field flipping, Yellowstone's geysers erupting catastrophically, and colossal solar flares damaging the earth. I remember Dylan admitting to me that out of fear he initially accepted the idea of consciousness shifting as a way to avert human disaster because, as he put it, "it's better than the end of the world." I agreed with Dylan's opinion that the articles he came across on the Internet focused on New Agers, astrology, and cosmic telepathy "because people think it's more interesting than true, mundane facts." But then my eyes brightened when he told me that when it comes to the Maya, "the facts are already fascinating!" You were right, Dylan. The ancient Maya do not need us to dress up their culture in the garb of Western ideas.

We may be attracted to the seeming holistic nature of Maya cosmology, but when we look carefully and critically at *all* the evidence, instead of only the pieces that suit our own particular fancy, we discover that the Maya had more important things to think about than seeking meaning in a "center of the universe" thousands of light-years away. For them the center was right *here*—in "all the sky earth," as the *Popol Vuh* tells it. Literal-minded Y12 prophets saturate the ancient Maya record with techno-scientific terminology and translate their creation myths into quantified analyses. Their precise predictions emerge as a cultural fashion of our times. It has all happened before, as I have shown. American pop culture today is preoccupied with cataclysmic endings as much as it is with new beginnings—with being redeemed. No wonder we love stories about people and sports teams who have been counted out and then somehow "come back from the dead."

Am I so culturally self-centered to think that the Maya message of 2012 is really intended for me? Do I believe the Maya tapped into a cosmic pipeline leading to lost knowledge and superior wisdom that could save the world? Will December 21, 2012, suddenly usher in a universal moment of the dawning of a new collective consciousness? At a workshop on Maya hieroglyphic writing that I attended recently, I asked a professed Maya shaman in attendance, "What can we expect in 2012 when your Long Count calendar overturns?" His answer gave me some hope that maybe all the Y12 hype might do us all some good. He replied: "Only the cycle will end. Time will continue, and we will learn to live in peace and harmony, for we are all a part of a plan to help the gods complete the creation and perfection of the world." Was he a "card-carrying" shaman? I cannot confirm the authority with which this Maya man spoke, but I definitely approved of his message. All of us have a role to play. The life of our planet has always depended on us and the actions we take while we are here on earth. We do not need superior mathematical knowledge, cosmic alignments, precise timings, or even God(s) to make that kind of prediction.

# NOTES

## CHAPTER 1:
## INTRODUCTION: HOW DYLAN GOT ME STARTED

1. L. Joseph, *Apocalypse 2012: An Investigation into Civilization's End* (New York: Broadway, 2008), back cover.

## CHAPTER 2: WHAT'S IN STORE?
## A USER'S GUIDE TO 2012 MAYA PROPHECIES

1. E. Todras-Whitehill, "Touring the Spirit World," *New York Times*, April 29, 2007, Travel section, 7.

2. J. Rivard, "A Hierophany at Chichen Itza," *Katunob* 7:3 (1970): 51–55.

3. R. Grant, *Gnosticism: A Sourcebook of Heretical Writings from the Early Christian Period* (New York: Harper, 1961), 18.

4. G. Stray, "Beyond 2012: Catastrophe or Ecstasy; A Complete Guide to End of Time Predictions," www.diagnosis2012.co.uk (accessed March 10, 2008).

5. Ibid. See also T. and D. McKenna, *The Invisible Landscape: Mind, Hallucinogens, and the I Ching* (New York: Seabury Press, 1975).

6. If DNA is a sentence and the nucleotides are the operations, then the codons are the words. I owe that little analogy to my colleague, biologist Frank Frey.

7. J. Argüelles, *The Mayan Factor: Path Beyond Technology* (Santa Fe, NM: Bear and Co., 1987), 184.

8. Ibid., 170.

9. Quetzalcoatl (Kukulcan to the Maya) is also the Mesoamerican god of creativity. His mischievous, darker counterpart Tezcatlipoca is often misinterpreted as the devil. The black and white Christian-duality of evil-good, however, does not accurately represent Maya beliefs.

10. J. M. Jenkins, *Maya Cosmogenesis 2012: The True Meaning of the Maya Calendar End Date* (Santa Fe, NM: Bear and Co., 1998), xxxii.

11. It is customary to refer to the Milky Way Galaxy, the vast system of 200 billion stars, of which the sun is but one, and the accompanying interstellar dust and gas, with a capital *G*. Astronomers reserve the lower-case *g* for the billions of other known systems that lie at vast distances.

12. Izapa is not a Maya site but rather a site occupied my Mixe-Zoque speakers.

13. Modern astronomy identifies the center of the Milky Way Galaxy as a point located in Sagittarius, some 25,000 light-years from the position of the sun and its system of planets. Thus, we dwell in a kind of "galactic suburbia."

14. Quiché is one of some twenty-nine dialects spoken by the contemporary Maya, who live in the region of the western Guatemalan highlands.

15. Jenkins, *Maya Cosmogenesis 2012*, 107.

16. Ibid., 150.

17. Ibid., 311.

18. Ibid., appendix 5.

19. Accessed May 10. 2009.

20. C. J. Calleman, *Solving the Greatest Mystery of Our Time: The Maya Calendar* (London: Garev Publishing International, 2000).

21. www.calleman.com (accessed March 25, 2008).

22. See, for example, C. Zaleski, *Otherworld Journeys: Accounts of Near Death Experience in Medieval and Modern Times* (New York: Oxford University Press, 1987).

23. D. Pinchbeck, *2012: The Return of Quetzalcoatl* (New York: Jeremy Tarcher / Penguin, 2006), 307–308.

24. Ibid., 308.

25. Ibid., 309.

26. Joseph, *Apocalypse 2012*, back cover.

27. Ibid., 114.

28. Ibid., 105.

29. Jenkins, *Maya Cosmogenesis 2012*, 331.

30. Argüelles, *The Mayan Factor*, 15.

31. www.greggbraden.com (accessed March 20, 2008).

32. "Mayan Perspectives on 2012," www.stetson.edu/~rsitler/perspectives (accessed May 17, 2008). Also see R. Sitler, "The 2012 Phenomenon: New Age Appropriation of an Ancient Maya Calendar," *Novo Religio: The Journal of Alternative and Emergent Religions* 9:3 (2006): 24–38, esp. 28.

33. Don Alejandro Cirilo Perez Oxlaj, head of the National Mayan Council of Elders of Guatemala, in his address at the inaugural of Guatemalan President Colom on January 2008.

34. José María Tol Chan, K'iche' daykeeper from Chichicastenango (2006).

## CHAPTER 3: WHAT WE KNOW ABOUT THE MAYA AND THEIR IDEAS ABOUT CREATION

1. T. Knowlton, *Words and Worlds of Classical Yucatecan Maya Creation Myths* (Boulder: University Press of Colorado, forthcoming).

2. See, for example, G. Vail and A. Aveni, *The Madrid Codex: New Approaches to Understanding an Ancient Maya Manuscript* (Boulder: University Press of Colorado, 2004).

3. A. Tozzer, *Landa's Relación de las Cosas de Yucatan, a Translation* (Cambridge, MA: Peabody Museum of Archaeology and Ethnology, Harvard University, 1941), 27

4. D. Tedlock, *Popol Vuh: The Definitive Edition of the Mayan Book of the Dawn of Life and the Glories of Gods and Kings*, rev. ed. (New York: Simon and Schuster, 1996), 146.

5. See, for example, D. Webster, *The Fall of the Ancient Maya: Solving the Mystery of the Maya Collapse* (London: Thames and Hudson, 2002);

and A. Demarest, D. Rice, and P. Rice, eds., *The Terminal Classic in the Maya Lowlands* (Boulder: University Press of Colorado, 2004).

6. T. Stanton and A. Magnoni, eds., *Ruins of the Past: The Use and Perception of Abandoned Structures in the Maya Lowlands* (Boulder: University Press of Colorado, 2008).

7. J. L. Stephens, *Incidents of Travel in Central America, Chiapas, and Yucatan* (New York: Harper and Brothers, 1841), 1:159–160.

8. R. Wauchope, *Lost Tribes and Sunken Continents: Myth and Method in the Study of American Indians* (Chicago: University of Chicago Press, 1962), 16.

9. L. Schele and P. Mathews, *The Code of Kings: The Language of Seven Sacred Maya Temples and Tombs* (New York: Scribner, 1998), 161–164. For more on Maya time units, see Chapter 4.

10. M. Edmonson, *The Ancient Future of the Itza: The Book of Chilam Balam of Tizimin* (Austin: University of Texas Press, 1982), 5, 6–7, 8.

11. M. Coe, *Breaking the Maya Code* (New York: Thames and Hudson, 1992); M. Miller, *The Art of Mesoamerica from Olmec to Aztec* (New York: Thames and Hudson, 1986); S. Martin and N. Grube, *Chronicle of Maya Kings and Queens* (London: Thames and Hudson, 2000).

12. L. Schele, *Notebook for the XXIst Maya Hieroglyphic Forum at Texas* (Austin: University of Texas Press, 1997), 199.

13. Ibid., 199; lines 1 and 2 are from the *Book of Chilam Balam of Tizimin*, 14v, and lines 3 and 4 are from the *Book of Chilam Balam of Chumayel*, 43, as translated and interpreted by Knowlton in *Words and Worlds of Classical Yucatecan Maya Creation Myths*.

14. M. Looper, *Lightning Warrior: Maya Art and Kingship at Quirigua* (Austin: University of Texas Press, 2003), 159.

15. I am indebted to Susan Milbrath for sharing with me her opinions on the extreme unlikelihood that the text of Tortuguero 6 has anything to do with predictions of cataclysmic events. For a colorful résumé of 2012 evidence in the inscriptions, see Mark Van Stone, *It's Not the End of the World: What the Ancient Maya Tells Us about 2012* (Crystal River, FL: FAMSI, 2008) and *2012: Science and Prophecy of the Ancient Maya* (forthcoming).

16. Tedlock, *Popol Vuh*, 77.

17. A. Aveni and H. Hartung, "Water, Mountain, Sky: The Evolution of Site Orientations in Southeastern Mesoamerica," in *Precious Green Stone Precious Feather: In Chalchihuitl In Quetzalli; Mesoamerican Studies*

*in Honor of Doris Heyden*, ed. E. Quiñones Keber (Lancaster, CA: Labyrinthos, 2000), 122.

18. Ibid., table 1.

19. A. Aveni, A. Dowd, and B. Vining, "Maya Calendar Reform: Evidence from Orientations of Specialized Architectural Assemblages," *Latin American Antiquity* 14:2 (2003): 159–178.

20. J. Guernsey, *Ritual and Power in Stone: The Performance of Rulership in Mesoamerican Izapan Style Art* (Austin: University of Texas Press, 2006). For further discussion of Stela 25 imagery, see also C. Coggins, "Creation Religion and the Numbers at Teotihuacan and Oaxaca," *RES* 29–20 (1996): 19–38.

21. M. Looper and J. G. Kappelman, "The Cosmic Umbilicus in Mesoamerica: A Floral Metaphor for the Source of Life," *Journal of Latin American Lore* 21:1 (2001): 3–54, esp. 16.

22. S. Milbrath, in *Star Gods of the Maya: Astronomy in Art, Folklore, and Calendars* (Austin: University of Texas Press, 1999), 288–291, offers the most thorough discussion of Maya references to the Milky Way.

23. D. Freidel, L. Schele, and J. Parker, *Maya Cosmos: Three Thousand Years on the Shaman's Path* (New York: William Morrow, 1993), esp. 76, 90. When I reviewed the book, I characterized it as a "curious combination of reasoned scholarship and emotional revelation," noting that "what will stand up under close scrutiny remains to be seen." See A. Aveni, "Review of Maya Cosmos," *American Anthropologist* 98 (1996): 197–198. I stick by that statement today.

24. Milbrath, *Star Gods of the Maya*, 288–291.

25. See the chapter on "Space, Are Maps Really Territory?" in my book *Uncommon Sense: Understanding Nature's Truths across Time and Culture* (Boulder, University Press of Colorado, 2006).

26. See, for example, D. Carrasco and S. Sessions, eds., *Cave, City and Eagle's Nest: An Interpretive Journey through the Mapa de Cuauhtinchan No. 2* (Albuquerque: University of New Mexico Press, 2007).

27. G. Hawkins in collaboration with John B. White, *Stonehenge Decoded* (Garden City: Doubleday, 1965).

28. J. Hawkes, "God in the Machine," *Antiquity* 41 (1967): 174.

29. D. Ulansey, *The Origins of the Mithraic Mysteries: Cosmology and Salvation in the Ancient World* (New York: Oxford University Press, 1989).

30. R. Bauval and A. Gilbert, *The Orion Mystery: Unlocking the Secrets of the Pyramids* (New York: Crown, 1994).

## CHAPTER 4: THE CALENDAR:
## JEWEL OF THE MAYA CROWN

1. Historical evidence directly connects the Aztec reckoning of the 260-day count with the Venus visibility period. E. O'Gorman, ed., *Fray Toribio Motolinia, El Libro Perdido* (Mexico City: Consejo Nacional para los Cultura y las Artes, 1989), 83–85.

2. B. Tedlock, *Time and the Highland Maya*, rev. ed. (Albuquerque: University of New Mexico Press, 1982), 155–156.

3. Tozzer, *Landa's Relación de las Cosas de Yucatan*, 169.

4. B. de Sahagún, *Florentine Codex: General History of the Things of New Spain*, trans. C. Dibble and A. Anderson (Santa Fe, NM: School of American Research; and Ogden: University of Utah Press, 1979), 4–5: 143.

5. Ibid., 7: 25.

6. For a discussion of how the Maya and Christian calendars are correlated, see A. Aveni, *Skywatchers* (Austin: University of Texas Press, 2001), 207–210. Although the correlation question was a contentious problem until the late twentieth century, an overwhelming number of Mayanists now believe that either of the two forms of the so-called Modified Thompson Correlation (they differ by two days) is the most consistent with historical and astronomical data.

7. Ibid.

8. P. Rice, *Maya Calendar Origins: Monuments, Mythistory, and the Materialization of Time* (Austin: University of Texas Press, 2007).

9. Edmonson, *The Ancient Future of the Itza*, 121.

## CHAPTER 5: THE ASTRONOMY BEHIND
## THE CURRENT MAYA CREATION

1. See, for example, Milbrath, *Star Gods of the Maya*, 203.

2. See my essay, "Where Orbits Came from and How the Greeks Unstacked the Deck," in *Uncommon Sense: Understanding Nature's Truths across Time and Culture*, by A. Aveni, 25–42 (Boulder: University Press of Colorado, 2006).

3. The rate of precession, in seconds of arc, is given by $P = 50.''2564 + 0.222T$, where $T$ is measured in centuries of tropical years (365d.2422) forward $(+)$ / backward $(-)$ from AD 1900.

4. Basically, it takes the sun 365.25 days minus 11 minutes to make an equinox-to-equinox circuit (one tropical year). It takes the sun 365.25 days plus 9 minutes to make a star-to-star (one sidereal year) circuit: therefore, the equinox must be moving relative to the stars.

5. S. Kabata, S. Sugiyama, A. Aveni, and T. Murakami, "Pyramid of the Moon: Preliminary Interpretation of New Data," paper presented at 66th Society of American Archaeologists Meeting, New Orleans, April 18–21, 2001.

6. I. Šprajc, "Astronomical Alignments at Teotihuacan, Mexico," *Latin American Antiquity* 11:4 (2000): 403–415.

7. Aveni, *Skywatchers*, 108–113.

8. Aveni, *Skywatchers of Ancient Mexico* (Austin: University of Texas Press, 1980), 103.

9. B. MacLeod, "The 3-11-Pik Formula" (unpublished manuscript, privately circulated, March 2008).

10. Jenkins, *Maya Cosmogenesis 2012*, Appendix 3 and 45, 74.

11. M. Grofe, "The Serpent Series: Precession in the Maya Dresden Codex" (Ph.D. dissertation, Native American Studies, University of California, Davis, 2007).

12. H. Bricker and V. Bricker, *Astronomy in the Maya Codices* (Philadelphia: American Philosophical Society, in press).

13. I owe this suggestion to John Justeson, personal communication, July 10, 2008.

14. This result differs from the one John Major Jenkins gives (*Maya Cosmogenesis 2012*, 113–114) and which he insists the Maya calculated, although he does not say how. J. Meeus, in *Mathematical Astronomy Morsels* (Richmond, VA: Willmann-Bell, 1997), 301–303, has calculated that the crossing point of the galactic equator and the solstices occurred in May 1998. His calculations also show that this alignment was within two-thirds of a degree, or a little more than the width of the disk of the sun, of that mark within a fifty-year period surrounding that date. This is a very small tolerance for naked-eye observations of such a phenomenon.

15. D. Raup and J. Sepkoski Jr., "Periodicity of Extinctions in the Geologic Past," *Proceedings of the National Academy of Sciences* 81 (1984): 801–805.

16. For a discussion, see Aveni, *Skywatchers*, 117–118.

17. Venus is the third strongest tide raiser, producing a wave averaging 1/500 inch in height. Interestingly, the tide-raising force varies

according to the inverse cube of the distance between the attracting bodies.

## CHAPTER 6:
## WHAT GOES AROUND: OTHER ENDS OF TIME

1. R. Frazer, ed., *The Poems of Hesiod* (Norman: University of Oklahoma Press, 1983), 103.

2. Ibid.

3. N. Campion, *The Great Year: Astrology, Millenarianism, and History in the Western Tradition* (London: Arkana, 1994), 1.

4. Acts 8:9–11.

5. Ibid., 21–22.

6. J. Kirsch, *A History of the End of the World: How the Most Controversial Book in the Bible Changed the Course of Western Civilization* (San Francisco: Harper San Francisco, 2006).

7. H. Schwartz, *Century's End: An Orientation Manual for the Year 2000*, rev. ed. (New York: Barnes and Noble, 1999), 6.

8. 2 Peter 3:8.

9. Lactantius, *Divine Precepts*, quoted in S. J. Gould, *Questioning the Millennium: A Rationalist's Guide to a Precisely Arbitrary Countdown* (New York: Harmony Books, 1997), 75–76.

10. Ibid.

11. Schwartz, *Century's End*, 148.

12. Virgil, *Aeneid*, vii, 44.

## CHAPTER 7: ONLY IN AMERICA

1. Virgil, *Aeneid*. The words on the U.S. dollar indeed come from antiquity.

2. T. Paine, *The Rights of Man,* ed. H. Collins (Harmondsworth, Middlesex: Penguin, 1969), 193.

3. Revelation 21:2, 16, 18.

4. D. Thompson, *The End of Time: Faith and Fear in the Shadow of Millennium* (London: Sinclair-Stevenson, 1996), 96.

5. S. Stein, *Encyclopedia of Apocalypticism*, vol. 3: *Apocalypticism in the Modern Period and the Contemporary Age* (New York, Continuum International Publishing Group, 2000), 52.

6. P. Boyer, *When Time Shall Be No More: Prophecy Belief in Modern American Culture* (Cambridge, MA: Belknap Press of Harvard University Press, 1992), 70.

7. I am grateful to my colleague in Religious Studies, Christopher Vecsey, for initially directing me to this information, as well as for valuable discussions centered on interpreting it in the context of 2012. I also drew from Jonathan Kirsch's work in preparing this section (Kirsch, *A History of the End of the World*).

8. A. Aveni, *Behind the Crystal Ball: Magic, Science, and the Occult from Antiquity through the New Age* (New York: Times Books, 1996; reprinted by University Press of Colorado, Boulder, 2002), chapter 17.

9. F. Podmore, *Mediums of the Nineteenth Century* (New Hyde Park, NY: University Books, 1963 [1902]), 1: 306.

10. C. G. Harrison, *The Transcendental Universe: Six Lectures on Occult Science, Theosophy, and the Catholic Faith; Delivered before the Berean Society* (Hudson, NY: Lindisfarne Press, 1993), 209. I owe a debt of gratitude to my colleague Joscelyn Godwin, both for resource material and helpful discussion of this episode.

11. R. Guénon, *The Reign of Quantity and the Signs of the Times* (London: Luzac, 1953), 310.

12. Ibid., 312.

13. Daniel 8:14.

14. Ibid., 18–27.

15. Daniel 7:18–27.

16. Kirsch, *A History of the End of the World*, esp. chapter 2.

17. R. Numbers, *The Creationists: From Scientific Creationism to Intelligent Design*, expanded ed. (Cambridge, MA: Harvard University Press, 2006), chapter 18.

18. C. P. Smyth, *Our Inheritance in the Great Pyramid* (New York: Bell, 1990), 551. The arithmetical error is his, not mine.

19. F. Waters, *Mexico Mystique: The Coming of the Sixth World of Consciousness* (Chicago: Sage Books, 1975), 163.

20. Letter to the editor in *The Pyramid Guide*, quoted in Aveni, *Behind the Crystal Ball*, 253.

21. P. Tompkins, *Mysteries of the Mexican Pyramids* (New York: Harper & Row, 1976), 389.

22. L. Arochi, *La Piramide de Kukulacan: Su Simbolismo* (Mexico City: Panorama, 1987).

23. Coe, *Breaking the Maya Code*.

24. These secrets were supposed to answer the questions, What does hell really look like? How can a sinner's soul be saved? and When will the Pope die?

25. G. Halsell, *Prophecy and Politics: Militant Evangelists on the Road to Nuclear War* (Westport, CT: Lawrence Hill and Company, 1986), 45. See Kirsch's excellent chapter on "The Godless Apocalypse" (in *A History of the End of the World*) for a thorough discussion of secular apocalypticism.

26. D. Nussbaum Cohen, www.jewishsf.com/bk990122/usafalwell. htm (accessed November 10, 2007).

27. G. de Santillana and H. von Dechend, *Hamlet's Mill: An Essay Investigating the Origins of Human Knowledge and Its Transmission through Myth* (Boston: Gambit, 1969), cover.

28. G. Michanowsky, *The Once and Future Star* (New York: Hawthorn Books, 1977), 47.

29. Ibid., 77–78.

30. Ibid.

31. Bauval and Gilbert, *The Orion Mystery*.

32. R. Bauval, *The Egypt Code* (London: Arrow, 2007), xxiii–xxiv.

33. Ibid., xviii.

34. Ulansey, *The Origins of the Mithraic Mysteries*.

35. R. L. Gordon, "The Sacred Geography of a Mithraeum: The Example of Sette Sfere," *Journal of Mithraic Studies* 1:2 (1976): 119–165.

36. W. Sullivan, *The Secret of the Incas: Myth, Astronomy, and the War against Time* (New York: Crown, 1996).

37. Waters, *Mexico Mystique*, 163.

38. Waters may have acquired his erroneous date from anthropologist Michael Coe's first edition of *The Maya* (1966), 149. Coe states that Armageddon would overtake degenerate people of the world on December 24, 2011, when the great cycle and the Long Count reaches completion.

39. Santillana and von Dechend, *Hamlet's Mill*, cover.

40. Schwartz, *Century's End*, 223. See also N. Cohn, *The Pursuit of the Millennium* (Fairlawn, NJ: Essential Books, 1957).

41. Jenkins, *Maya Cosmogenesis 2012*, 332.

42. Ibid., 200.

43. Jenkins, *Maya Cosmogenesis 2012*.

44. F. Drake and D. Sobel, *Is Anyone Out There? The Scientific Search for Extraterrestrial Intelligence* (New York: Delacorte Press, 1992), 160.

45. Argüelles, *The Mayan Factor*, 194.

46. B. Vranich, personal communication, December 30, 2007.

47. S. Jacoby, *The Age of American Unreason* (New York: Random House, 2008), convincingly documents the case.

# GLOSSARY

**apocalypse**. From the Book of Revelation, a collection of divine prophecies related to the end of time preceding the Second Coming of Christ; also a general term applied to the disclosing, to certain privileged persons, of secret information affecting humanity.

**autumnal equinox**. See **equinox**.

**Big Bang**. The cosmological model, best supported by current evidence, that theorizes that the universe was created in a singular flash some 13.7 billion years ago.

**black hole**. A region of space in which the gravitational pull is so powerful that nothing, including electromagnetic radiation, can escape it.

**celestial equator**. The prolongation of the earth's equator, or plane of rotation, onto the sky. This great circle is 90 degrees distant from the **celestial poles**.

**celestial poles**. The points marking the extension of the poles of rotation of the earth onto the sky.

**commensuration**. The property whereby a quantity can be related to another quantity by a ratio of two small whole numbers; for example, because five Venus years of 584 days are equivalent to eight seasonal years, counted as 365 days, we say that the two periods are commensurable in the ratio of 5 to 8.

**conjunction**. Any lineup of celestial bodies, such as the sun and the moon in its new phase.

**ecliptic**. The extension onto the sky of the earth's plane of revolution about the sun, a great circle making an angle of 23.5 degrees with the **celestial equator**. As far as terrestrial observers are concerned, this circle traces the annual motion of the sun on the sky relative to the background of distant stars (the **zodiac**).

**equinox**. One of two points on the celestial sphere at which the sun, in its annual course along the ecliptic, crosses the celestial equator. The **vernal equinox** is the point of intersection of the **ecliptic** and the **celestial equator** where the sun passes from the southern to the northern hemisphere. The autumnal equinox is the opposing intersection point where the sun passes from north to south. The equinox dates are approximately March 21 and September 22, respectively.

**galactic center**. The center of rotation of the Milky Way Galaxy. About 25,000 light-years distant, it is located in the constellation of Sagittarius.

**galactic equator**. A great circle in the sky that marks the central line of the Milky Way Galaxy. It is defined by the peak concentration of neutral hydrogen observed at a radio wavelength of twenty-one centimeters. Also termed the "plane of the galaxy."

**galaxy**. A large star system, generally of flattened form, held together by gravity.

**Gnosticism**. A form of religious internationalism in which knowledge of God is thought to be revealed experientially and personally via a series of hierarchically ordered emanations that lead to the highest "One."

**Great Rift**. A group of dark interstellar clouds along the Milky Way stretching from Cygnus to Sagittarius.

*haab*. The Maya **seasonal year** cycle of 365 days, reckoned without leap years, and organized by eighteen months each of twenty days, to which an end month of five days is added.

**heliacal rising.** The first appearance of a star (or planet) after its invisibility because of **conjunction** with the sun.

**heliacal setting.** The last appearance of a star (or planet) before its invisibility because of **conjunction** with the sun.

**Long Count.** The largest base-20 calendrical cycle once used by several Mesoamerican cultures, most notably the Maya. The cycle comprises the following units: *k'ins* (days), *uinals* ("months" of twenty days), *tuns* (years of eighteen *uinals*), *katuns* (scores of *tuns*), and *baktuns* (thirteen *katuns*).

**Mayanism.** A collection of New Age beliefs centered around the notion that ancient Maya wisdom contains higher knowledge ranging from the nineteenth- and early twentieth-century belief that they possessed secrets of lost Atlantis to the widely held contemporary view involving extraterrestrials.

**Metonic cycle.** Devised by the fifth-century BC Greek astronomer Meton, the time it takes to return a given phase of the moon to the same date in the seasonal year; 235 months, or nineteen years.

**Milky Way.** The **galaxy** in which the solar system, along with 200 billion other stars, is situated. Seen from the eccentric position of the earth, which is situated close to its plane and about two-thirds of the way between the galactic center and the outer edge of the disk, it appears as a faint luminous band of light about 15 degrees (thirty moon diameters) in average width that stretches all the way around the sky.

**Millennialism.** A doctrine of medieval origin purporting that in successive thousand-year periods the world will end in cataclysm and, upon the destruction of evil, be reborn.

**precession of the equinoxes.** The slow conical motion of the earth's axis of rotation about the poles of the **ecliptic**, resulting in a motion of the **celestial poles** and **equinoxes** among the stars in a cycle of approximately 26,000 years.

**Rapture.** The prophetic theory that teaches that prior to the battle of Armageddon, Christ will return in stages.

**seasonal year.** See **tropical year.**

**sidereal period.** The interval between successive passages of a body by a given star; for example, for the moon (the sidereal month), 27.32166 days.

**solar zenith passage**. Applicable only to the tropics (i.e., between latitudes 23.5°N and 23.5°S) the two dates of the tropical year when the sun crosses the overhead position.

**solstices**. See **summer solstice** and **winter solstice**.

**summer solstice**. The point on the celestial sphere where the sun reaches its greatest distance north of the **celestial equator**; about June 21.

**tropical year**. The period of revolution of the earth about the sun (or, as we see it, the time it takes the sun to successively pass the vernal equinox, 365.24220 days). More commonly known as the **seasonal year**.

**tzolkin**. The Maya cyclic "count of days" (260) achieved through the combination of thirteen numerals and twenty day names (*uinals*).

**vernal equinox**. See **equinox**.

**winter solstice**. The point on the celestial sphere where the sun reaches its greatest distance south of the celestial equator; about December 21.

**world ages**. The theory that all human history is divided into time units that are multiples of 1,000 years; thus, the Six Ages of the World according to St. Augustine are divided into 2,000-year periods, from the birth of Adam to Revelation and the Apocalypse.

**zenith**. The point on the sky directly overhead (opposite the direction of a plumb line).

**zodiac**. The twelve constellations, or, in astrology, "signs" that lie along the plane of the **ecliptic**; they make up a band of 16 degrees average width.

# INDEX

*Page numbers in italics indicate illustrations.*